Mechanical Engineering Practices in Industry

Mechanical Engineering Practices in Industry

A Beginner's Guide

Dhruba J Syam

First edition published 2023
by CRC Press
4 Park Square, Milton Park, Abingdon, Oxon, OX14 4RN

and by CRC Press
6000 Broken Sound Parkway NW, Suite 300, Boca Raton, FL 33487-2742

CRC Press is an imprint of Informa UK Limited

Print edition not for sale in South Asia (India, Sri Lanka, Nepal, Bangladesh, Pakistan or Bhutan)

ISBN: 9781032516103 (hbk)
ISBN: 9781032516127 (pbk)
ISBN: 9781003403104 (ebk)

DOI: 10.4324/9781003403104

Typeset in Times New Roman
by Manakin Press, Delhi

Dedicated to the young Mechanical Engineers
who aspire to make a career in
Engineering Industries.

Author

Acknowledment

I express my sincere thanks to Sri. T.K. Banerjee, Ex-General Manager (Turbine Production) and Sri. Pavan Kumar Arora, Addl. General Manager (CNC Technology), both of BHEL Hardwar, for going through the manuscript and giving valuable suggestions for improvement/enrichment of the contents.

I am very much thankful to Dr. S.P. Harsha, Professor of the Mechanical Engineering at IIT-Roorkee, for his kind gesture of giving a Recommendatory FOREWORD for the book.

I am also thankful to M/s. MANAKIN PRESS, New Delhi, for accepting and undertaking the printing and publishing of the book.

Author

Foreword

"It was a great pleasure reading through the manuscript of book titled "Mechanical Engineering Practices in Industry: A Beginners Guide" authored by eminent Shri Dhruba J. Syam. Written in a masterly manner, it is a description of the basics and current scenario of mechanical engineering related practices. The topics dealt within the book will be good help to young mechanical engineers to have insight of industry working systems. Though there is a continuous progress from the aspects of machine/s or component/s design to maintenance practices, but it requires basic understanding to a beginner and this effort fulfils that. Shri Syam has touched upon all issues related to modern practices in industry which makes his book an invaluable text for graduating engineers particularly in mechanical engineering disciplines as well as all those who are concerned with maintenance practices."

Dr. S.P. Harsha
MIED, IIT-Roorkee

Foreword

Detailed Contents

Preface

Industrial Enterprises engaged in manufacturing of engineering products such as consumer durables, machines, equipment, devices, structures etc. require trained engineers, technicians and skilled workers to run the day to day operations successfully and in a sustainable manner. In an Integrated Engineering Industrial Enterprise, the entire process of operation commonly comprise of design and engineering of products, manufacturing, marketing and after-sales-services, supported by related services like Quality Control, Materials Management, Financial Services, Plant Maintenance and Repair services, General Administration and such other functions which contribute towards its smooth running for turning out saleable products.

For most of the technical functional areas particularly for the shop-floor activities, **services of Mechanical Engineers become essential, since most of the shop floor activities are predominantly based on Mechanical Engineering Based Processes.**

In spite of the present IT-boom offering alternative employment opportunities, a large number of mechanical engineers may have to find jobs in the engineering goods manufacturing sector where substantial employment opportunities exist . Young mechanical engineers have to therefore, prepare themselves reasonably well to face the challenges of industrial working systems and work-environments.

During professional working for over 40 years with industrial and business houses, the Author had to deal with and guide/mentor a large number of young engineering and other business administration professionals (like MBAs, CAs) during their summer as well as during induction training in the corporate enterprises where the author served in managerial/senior management positions. **In the course of interaction with the young engineering professionals, the author had tried to identify and analyse the difficulties those young bright minds find themselves in when they join industrial enterprises to start their career and struggle to survive, not only in the job, but also, to a large extent, in the process of acquiring self-confidence in the beginning days, weeks and months, mainly due to lack of practice-oriented knowledge inputs.**

In the present era of competitive survival, seniors/bosses generally remain quite preoccupied to spare enough time and resources to MENTOR the young ones to the extent desirable, but at the same time they are not only rather choosy about selection but also quite demanding at the work place, expecting young engineers to perform with the required proficiency right from the beginning weeks, if not days, **with minimal mentoring**.

The four year undergraduate course (BE, B.Tech etc.) is quite loaded with theoretical contents and the students hardly find enough time and opportunity to adequately grasp the physical and practical aspects of application of various engineering theories that are being taught. It is only when these young engineers start working in the manufacturing industry tries to gradually pick-up working knowledge in bits and pieces. Therefore, certain practice-oriented knowledge inputs in the beginning days and months may help them acquire and enhance proficiency in the industrial working systems and processes.

The author in this book has attempted to provide certain practice-oriented knowledge inputs which may help young **mechanical engineers** who aspire to make a successful career in engineering goods manufacturing enterprises. The topics dealt with are expected to help broaden and enhance their knowledge base at an accelerated pace.

The book seeks to provide a combination of Engineering and Production/Manufacturing Management Aspects to enable young mechanical engineers to make a confident start at the work place, and eventually ascend to leading positions in the organization.

The Author sincerely hopes that young mechanical engineers will be benefitted by reading the book and assimilating the essence of the topics presented.

NOTE: Professional subjects like the present one deserve serious treatment in order to derive appreciable benefits of knowledge gathering to support career aspirations.

Author

1 Prologue

"I cannot teach anybody anything, I only make them Think".

—Socrates

1.0.

Industrial Enterprises manufacture various kinds of engineering and consumer products such as plants and machinery, equipment, structures, devices, consumer durables etc. for different sectors of economy and market segments; these enterprises also include process plants such as power generation stations, steel/metallurgical plants, petro-chemical plants, equipment and machinery for transportation and shipping, defence, space exploration, oil and gas exploration, agricultural implements etc.: the list can be endless.

Such industrial enterprises as mentioned above, generally follow certain shop-processes which **are predominantly mechanical engineering and technology oriented**. In fact, most of the activities that take place on the shop floors of manufacturing units are based on mechanical processes broadly related to the processes of **Shaping** (forming and machining), **Joining** (fabrication, welding) and **Finishing** (assembly, testing, surface treatment, packing, shipping), with the exception of certain specific processes relating to electrical and electronic in-process measurement and testing for products and/or components that work on the principles of electro-mechanical-electronic processes; even for such electrical and electronic products and components, barring the in-process measurements and testing as mentioned above, all shop floor processes are by and large are mechanical in nature.

1.1.

In view of the above, the first and the most preferred category of operating and managing engineers and technologists that are required to successfully run the industrial operations is the Mechanical Engineers/Technologists. In most of such industrial enterprises, the services of mechanical engineers are essential particularly where complex plants, machinery and equipment are installed and operated for productive outputs on a regular scale.

1.2.

Mechanical Engineering is a versatile and evergreen branch of engineering sciences. Mechanical Engineering branch encompasses a vast scope of activities and has many sub-branches to meet specific functional needs. A comprehensive detailing about these will be an enormous task and dealing with these in one book like this is not justifiably possible. Engineering Professionals, who aspire to work and make a career in the Engineering Goods Manufacturing Enterprises, the relevant topics, dealt with in a concise but practice-oriented presentation, in the Chapters that follow, may be useful for learning and developing professional skills and gaining knowledge that may be useful for facing the challenges of managerial/supervisory responsibilities and career growths.

1.3.

Young Mechanical Engineers who get employed in industrial units, may be posted in any of the various departments of the enterprise. While some of these departments/functional areas may predominantly have technical activities, others may deal with matters which are a mix of technical, commercial, administrative as well as liaison services.

1.4.

Besides having such techno-commercial-liaison assignments, engineers may also be required to deal with certain general management affairs relating to employees/subordinate staff, industrial-relation matters, dealing with customer services, liaison with external agencies like Factory Inspector, Boiler Inspectors, Explosive Inspectors, Pollution Control Authorities, Auditors, Labour Commissioners, other state and local bodies including, municipal, civil and police administrations, suppliers and contractors, other external service providers.

1.5.

Seniors/Management Functionaries will expect that the young engineers should be able to discharge the assigned responsibilities in a competent and matured manner to deliver the desired results and avoid creating bottlenecks/ problems in the operation due to lack of knowledge, proficiency and the desired level of competency. Such expectations from the bosses/management throws a challenge which must be rather successfully faced and handled to the best of ability, not only for making positive result oriented contributions (to justify salary and perks), but also to enhance career prospects; no self-respecting young intelligent person, particularly educated and trained professionals, would relish to be branded as sluggish and incompetent. This is a perpetual challenge before the young professionals and even the middle-management executives, who have to continuously strive for acquiring up to date knowledge and proficiency, but also make a positive impression with the seniors who are expected to evaluate their performance and decide matters relating to career advancement prospects.

1.6.

In today's highly competitive globalized business scenario, young engineering professional cannot afford to be lackadaisical and sluggish in their approach towards professional responsibilities since only ambition will not help; down to earth hard work coupled with a positive knowledge seeking frame of mind is also essential for achieving satisfactory level of competence for growth in career. There has to be a continuous **urge for knowledge gathering** (which is in fact, a life-time endeavour). **Job satisfaction** is another important factor in professional career. Once joining the enterprise, the young professionals can very rarely choose his/her assignment and almost never can choose his/her boss; but of course can choose certain skills and proficiency which have a demand in the market; but hopping from job to job without acquiring knowledge and competence, is like a rolling stone gathering no moss. Therefore, it is, to a large extent, barring some extreme and rare circumstances, as to how the young professional develops his/her attitude and work-ethics (and culture) so as to be successful in this assignments, at the same time get a reasonable level of job-satisfaction: **there is however, no universal standard/measuring scale for this.**

1.7.

Whatever may be the assignment allocated, no body normally prevents any enthusiastic young engineer to find out ways and means (within the disciplinary frame-work of the organization) to learn about things happening/being done in other departments, particularly those which have a bearing on the assigned functional responsibilities. Interacting with colleagues and friends working there as an when opportunities come by (striking an appropriate rapport at the work-place is a part of over-all learning process) will be beneficial; this can be in the form of interactive discussions, visits to the area of interest, reading relevant documents/books (many enterprises have library facilities- both central and departmental). In fact, most progressive organizations encourage young engineers to learn more and more about the assigned jobs including the relevant knowledge inputs from the other/associated departments. All enlightened progressive enterprises adopt active policies to enhance managerial skills in their young executives/supervisors through various HRD-Programmes.

1.8.

Referring to the 'Contents' pages of this book, attempts has been made to deal with various relevant topics which an young mechanical engineer may find useful not only for successfully carrying out of his day to day technical assignments, but also getting certain knowledge inputs that help develop his managerial skills, which in turn enhance his career prospects.

1.9.

It is to be appreciated that each of the topics dealt with in this book has the potential for a vast scope of coverage and volumes can be written and have been written by learned experts on these; but here, since the purpose is to deal with such topics in a single handy ready-reckoner type reference book for young professional learners, the treatment had to be done in a limited coverage without making the book a voluminous one which may discourage young readers.

1.10.

It is also to be mentioned here that young mechanical engineers working in the engineering goods manufacturing industries must also strive to acquire a reasonable level of working knowledge regarding metallurgical, electrical/ electronics, instrumentation, and information/data processing (computerised working) technologies which are relevant/associated with mechanical technologies/processes. This is not only required for mutual role appreciation (and also the limitations) amongst various associated engineering groups, but this also helps in intergroup rapport for an unified result oriented work-culture. In an industrial situation, it is not unusual to see a mechanical engineer is called upon to deal with various related mechano-electrical/electronics and mechano-metallurgical as well as general administrative matters, even, circumstances demanding, certain civil engineering activities.

1.11.

The objective of this book is to provide to Young Mechanical Engineers certain practice-oriented knowledge inputs that may help in acquiring reasonable level of proficiency in Techno-Organizational Management aspects at an accelerated pace, which in turn may help enhancing their career prospects.

"Don't judge each day by the harvest you reap, But by the seeds you plant".

—R.L. Stevenson

■ ■ ■

2

Industrial Organizations: Operational Characteristics

(With particular reference to engineering goods manufacturing enterprises)

(An awareness brief)

"An Organization should always think about acquiring new business, or to expand the existing business". **—Chanakya**

Remark: In the modern day context, Young Engineers, beside technical proficiency, must also get reasonably familiarized with the nuance of Business/ Organizational Administration Processes for enhancing their competence for ascending to leadership positions.

2.1.

Mechanical Engineers working in industrial enterprises, on many occasions, depending on their place of posting, have to work like a **multi-tasking engineer**, dealing with mechanical and electrical, even civil engineering related activities. Therefore, a mechanical engineer in such a working situation has to consciously work for acquiring multi-tasking skills and versatility, rather than focussing on the narrow specialization; such an approach will also help in acquiring broad-based managerial/leadership capabilities which are required at the higher positions in the organization. **However, this is not to opine** that specialization has no importance; on the contrary, in certain functional areas, specialized services are essentially required; in such cases, multi-tasking capabilities may not be expected of the professionals concerned. At the same time, acquisition of multi-functional managerial capabilities is essential for ascending to leadership positions in future as and when opportunities come by.

2.2.

Since most shop activities are based on mechanical engineering oriented processes, engineers working in such shops, whatever may be their engineering discipline by educational background, they ultimately have to act and behave like a shop mechanical/production engineer for all practical purposes.

Nevertheless, all young engineering professionals working in industry should have some working knowledge on product-realization processes that go on in the enterprise since after all, engineering concepts get converted into technological efforts towards development of products, processes for product-realization, and related services for commercial exploitation, besides other direct and indirect benefits to the enterprise as well as to the society at large, through rendering support to the national economic development efforts.

2.3.

The basic **Operational Systems** in an industrial organization are depicted in **Diagram No.2.1**. Depending on the type and size of the enterprise (unitary organizations or multiple-unit company/corporation), the matching organizational structure is adopted. There are a variety of options; but with the passage of time and under changing business environments, the organizational structure may undergo changes/revamping from time to time to meet the changing needs. **An example of** the organizational setup in a complex industrial enterprise is presented at **Diagram No.2.2**.

2.4.

Organizational Aspects that the young Engineering Professionals must get familiarized with in respect of Industrial Enterprises:

2.4.1 Industrial Organization

(A) Before joining or on joining the industrial enterprise, it is advisable that professional beginners make conscious efforts to get acquainted with the features and character of the organization. This will help them to orient themselves with the organizational work environment, the prevailing work-culture, and the objectives for which the enterprise is operating. Since by the very nature of their work-assignments/ responsibilities, they will find it difficult to be very effective in their day-to-day performance if they remain aloof and ignorant about the character of operation of the enterprise .

(B) There are different types of business/industrial organizations and the young professionals are advised to get some knowledge about this subject by reading relevant books.

NOTE: While this book does not propose to go into the detailed scope of the subject (which itself is a vast one necessitating substantially elaborated treatment), nevertheless, for the benefit of the young professionals, certain aspects are mentioned below as an Awareness Brief.

(C) Engineering Professionals, working in different areas of activities, should also get conversant with the **following aspects** for increasing their organizational knowledge-base, enhance working competency and confidence:*

(i) Organizational structure of the main enterprise, preferably with a knowledge of names and credentials of the key personnel (owner or owning group, President/CMD, MD, Directors, various Unit and departmental/ functional heads);

(ii) Organizational set up at **the unit** where the incumbent is working and the names and credentials of the key functionaries;

(iii) Organizational set up of the **department/shop** where the incumbent is working and the names and credentials of the key functionaries;

(iv) Product lines/Business lines of the Main Organization and that of the Units;

(v) Market/clientele for the products made/sold and the services rendered (both domestic as well as the foreign, if any);

(vi) **Technological Processes** followed and the facilities available;

(vii) If possible, some knowledge/information about the competitors (their standings, product range, market share etc.) should also be gathered (to the extent feasible);

(viii) **Company's short term and long term business objectives** (in case the same are declared through brochures/handouts, annual reports etc).

(ix) Company's expansion/diversification objectives/plans (in case the same are declared).

****NOTE: Reading the Company Brochure/Annual Report of the Company may provide a good amount of information on this.***

2.4.2 Shop Organization

(i) Understand the product and get familiarized with the manufacturing processes followed; know about other related activities that take place in the shops to support production efforts.

(ii) Understand the meaning , importance and the role played by the **PPC Group** and the action-programme issued by the group; get familiar with meanings of the terms such as: Work-Order, WIP, finished goods, material introduction plans,, shop/machine-loading, machine and labour productivity, stage-wise inspection and quality control, **Quality Plan**, shop safety plans/instructions.

(iii) Be acquainted with the personnel working there and know their educational status and skills, capabilities and work-attitudes. **(This may have to be done rather discretely and tactfully).**

(iv) Take care of materials in sub-stores, their inventory, methods and records of receipts, issue, and replacements, replenishments, returns.

(v) Take care of Archive and Documents (drawings, process sheets, job cards, material lists, other related records) including updates; **take care for weeding out/removal of old scraped ones from the work-centres.**

(vi) Keep vigilance regarding safety to prevent/minimise accidents, and be aware of the security and safety requirements (including those for fire hazards/fire incidence and electrical hazards).

(vii) Manage scrap collection, segregation and disposal as per prescribed methods.

(viii) Take care of shop services which may include efforts for cleanliness, tidying up of shop arrangements, arranging materials and tools, shop accessories and small tools, upkeep of plant and machinery etc., proper house-keeping and upkeep of shop hygiene. Please try to implement **5S system of work-culture**.

NOTE: Please refer to Chapter 13 for more about 5S *System.*

(ix) Organize and supervise small repairs and troubleshooting; be aware of **Jishu Hozen'** work-culture. ***NOTE: Please refer to Chapter 13 for more about 'Jishu Hozen' work-culture.***

(x) Keep an eye on short-shelf-life items and their consumption schedules/ replenishment schedules.

(xi) Get familiarized with and Organize **Management Reporting System (MIRs)** as per the prescribed systems being practiced in the organization.

(xii) Keep liaison with designers, technologists, concerned material/ inventory control and purchase executives, shop floor quality-control personnel, maintenance personnel and try to maintain best of working relationship/rapport with them; their cooperation is very important for your own performance success.

(xiii) Keep liaison with **customer's representatives** and the Commercial Department, and be conscious/informed about delivery and quality commitments.

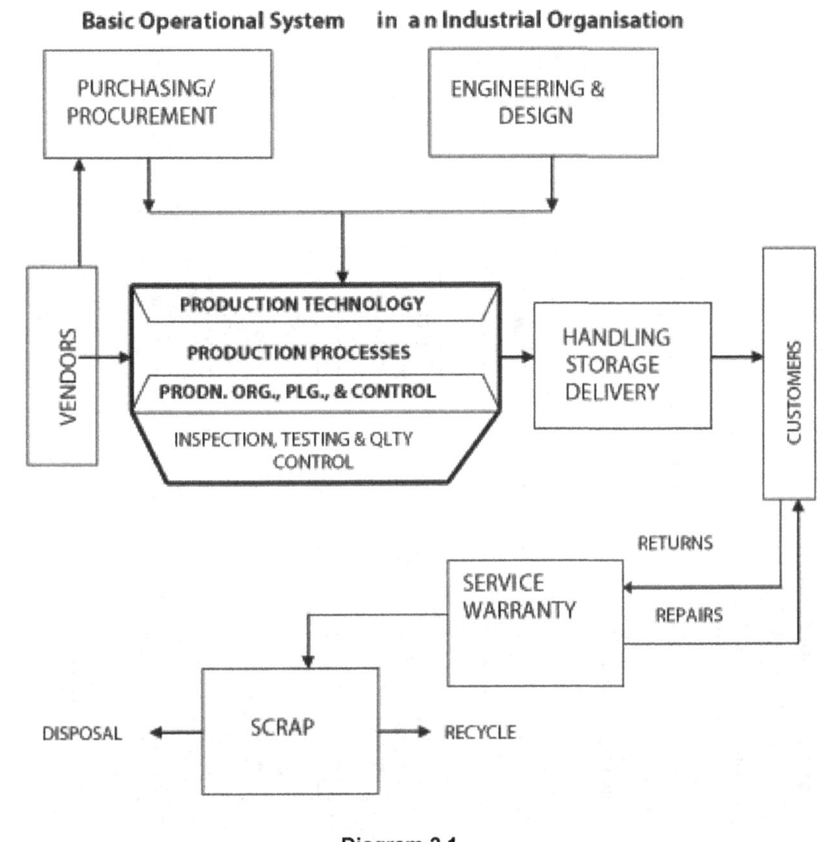

Diagram 2.1

(xiv) Be aware of statutory regulations and provisions/rules pertaining to shops, factory premises and labour force.

(xv) Be conscious about budgetary process and budgetary controls as practised in the company.

(xvi) Be aware of the prescribed **Delegation of Powers** and area of responsibilities assigned to you.

(xvii) Be conversant and conscious about costs and cost control aspects for your own work-centres.

(xviii) Be conversant with various company and statutory systems and procedures and take follow up action as prescribed.

(xix) Avoid remaining totally confined to own area of activity and be informed about other functions in the organization (isolation or seclusion can be detrimental for your own career growth as well as for your day to day performance, but be tactful (and discrete) in the approach.

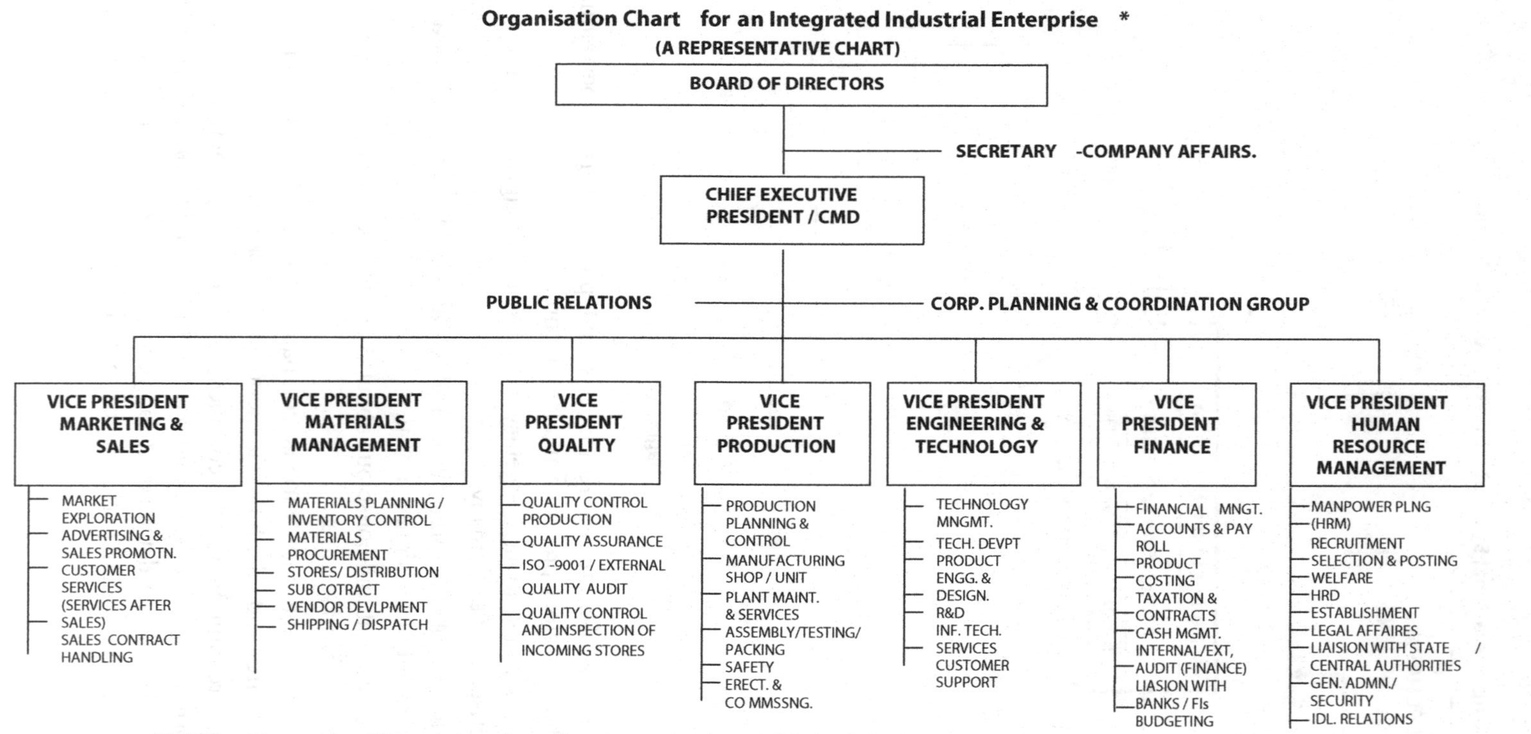
Organisation Chart for an Integrated Industrial Enterprise *
(A REPRESENTATIVE CHART)
BOARD OF DIRECTORS
SECRETARY -COMPANY AFFAIRS.
CHIEF EXECUTIVE PRESIDENT / CMD
PUBLIC RELATIONS
CORP. PLANNING & COORDINATION GROUP
VICE PRESIDENT MARKETING & SALES
MARKET EXPLORATION
ADVERTISING & SALES PROMOTN.
CUSTOMER SERVICES (SERVICES AFTER SALES)
SALES CONTRACT HANDLING
VICE PRESIDENT MATERIALS MANAGEMENT
MATERIALS PLANNING / INVENTORY CONTROL
MATERIALS PROCUREMENT
STORES/ DISTRIBUTION
SUB COTRACT
VENDOR DEVLPMENT
SHIPPING / DISPATCH
VICE PRESIDENT QUALITY
QUALITY CONTROL PRODUCTION
QUALITY ASSURANCE
ISO -9001 / EXTERNAL
QUALITY AUDIT
QUALITY CONTROL AND INSPECTION OF INCOMING STORES
VICE PRESIDENT PRODUCTION
PRODUCTION PLANNING & CONTROL
MANUFACTURING SHOP / UNIT
PLANT MAINT. & SERVICES
ASSEMBLY/TESTING/ PACKING
SAFETY
ERECT. & CO MMSSNG.
VICE PRESIDENT ENGINEERING & TECHNOLOGY
TECHNOLOGY MNGMT.
TECH. DEVPT
PRODUCT ENGG. & DESIGN.
R&D
INF. TECH. SERVICES
CUSTOMER SUPPORT
VICE PRESIDENT FINANCE
FINANCIAL MNGT.
ACCOUNTS & PAY ROLL
PRODUCT COSTING
TAXATION & CONTRACTS
CASH MGMT.
INTERNAL/EXT, AUDIT (FINANCE)
LIASION WITH BANKS / FIs
BUDGETING
VICE PRESIDENT HUMAN RESOURCE MANAGEMENT
MANPOWER PLNG (HRM)
RECRUITMENT
SELECTION & POSTING
WELFARE
HRD
ESTABLISHMENT
LEGAL AFFAIRES
LIAISION WITH STATE / CENTRAL AUTHORITIES
GEN. ADMN./ SECURITY
IDL. RELATIONS
*NOTE: - There can be different variants depending on operational size and the kind of Industrial Enterprise.

Diagram 2.2

(xx) **Be aware of the organizational hierarchy and the chain of Command** in the organization, as well as that in place in the functional area where posted.

(xxi) Take command of your assigned responsibility and carry the team with you; practise persuasive approach to the best extent possible.

(xxii) **Finally,** be positive and diligent in attitude to the best extent possible and practicable under the given circumstances; avoid conflicts and intrigues to the best extent possible.

2.5 INTER-FUNCTIONAL ROLE APPRECIATION AND MANAGERIAL CO-ORDINATION

Industrial Enterprises are multi-functional organizations. Although the main priority and thrusts are on **production related activities**, the other correlated functions and support services are also important for successful and smooth operations. Engineers must get acquainted with the scope of work of these correlated functions vis-à-vis the technical activities for enabling an unbiased role appreciation for mutual cooperation and benefits. **These correlated functions and support services in a composite industrial enterprise are more likely to be as listed under (major functional groups):**

(i) Materials Management: Procurement/purchase, storage, distribution, disposals, inventory control, conservation, shipment/transportation/logistics services, vendor development and liaison etc.

(ii) Commercial Management: Sales, Marketing, Order booking, advertising and publicity, customer services, customer contracts, after sales service coordination etc.

(iii) Finance & Accounting Functions: All matters relating to financial management and accounting functions such as sourcing of fund and cash/treasury management, budgeting and budgetary controls, realization of sales proceeds, liaison with banks, other financial institutions and authorities, internal auditing, allocation of funds, maintaining records of assets and liabilities of the company/enterprise, cash flow management, payroll matters, Profit & Loss Accounting, preparation of Balance Sheets, Coordination for external auditing, financial concurrence of transactions, contracts, dealing with loans/borrowings, **Letters of Credit, Bank Guarantees, etc.**

(iv) Plant Maintenance & Services Functions: This is basically a technical function which is different from manufacturing activities, but is an essential support service for smooth and trouble free productive functioning and upkeep of various plants and machinery, equipment, services, buildings/sheds and other infrastructural facilities, maintenance and repair of facilities, safety functions.

(NOTE: Mechanical Engineers will have a big role in this area of activities).

(v) **Personnel & Administrative Services (HRM, HRD functions):** General Administration, recruitment and posting of personnel, establishment matters, all aspects of **Human Resource Management** and **Human Resource Development** (training/re-training, skills development), industrial relations matters, security matters, liaison with external agencies like government/civil administration, police departments, local bodies; factory inspectors, labour department officials, also deals with disciplinary matters, employee welfare and amenities, and such other non-technical activities including Legal department.

(vi) **Engineering and Design function:** This is an important supporting technical function relating to product design/engineering, **product development** and **R&D**, rendering support services and guidance to shop technologists and shop technicians, engineering co-ordination amongst customers, production, commercial/marketing departments.

(vii) **Quality Assurance (QA) and Quality Control (QC) functions in conjunction with Product Engineering Group:** These are also supporting technical functions closely associated with production operations. While the Quality Assurance group, in conjunction with product engineering group, draw up conceptual technological plans and procedures for ensuring quality and reliability of products/components, the Quality Control group deals with actual shop floor activities for stage-wise inspection and in-process quality control checks by using various inspection tools and measuring devices as per the procedures and parameters prescribed by QA, the Customers and the designers.

(viii) **Public Relation**: This function may be either under Commercial Department or with the HRM department. This function may also deal with Publicity and Advertisement in conjunction with the Commercial/marketing group.

(ix) **Legal Department**: Dealing with all legal matters; in most organizations, the function is placed under the General Administration/HRM department. **At the company HQTRs, this cell may be attached to the Company Secretariat.**

(x) **External Erection-Commissioning Group:** This group may exist separately to take care of erection-commissioning of machines and equipment (produced and supplied to customers) at customer's premises/project sites. These group has to work in close conjunction with production and commercial departments and also with the representatives of the customers.

2.6 INDUSTRIAL RELATION MATTERS

The conduct of affairs pertaining to the day to day working relationship between the owners / management and the employees (and their representative forums like Trade Unions, Executive Association, Supervisors Association etc.), is generally perceived as the activities and processes that relate to the term **'Industrial Relation'**. However, in broader view, the scope of **'Industrial Relation'** may go beyond the traditional concept of employer – employee interactive relationship, and may also encompass the working relationships even with other externally associated agencies like customers, shareholders, suppliers, contractors, service providers, banks/FIs, Auditors, Local Bodies, Civil and Police Authorities, and such other state agencies: these agencies may have considerable direct and indirect influences and impacts on the running of operations of an **Industrial Enterprise**. Therefore, while framing the industrial relations policies and procedures, these external factors may also have to be kept in view, more so when the industrial units are operating in an era of globalized competitive market situation. Under such circumstances, the industrial enterprise will have to have a reasonably good reputation besides creditable operational performance. The perception that such external agencies can be dealt with through the process of **'Public Relation'** exercises, is not very logical, since public relation process predominantly deal with image-building exercises which may or may not address the various issues with respect to the working / business relationship with these external agencies as mentioned above. There is, however, a difference in strategies to be adopted in case of dealing with employee – specific affairs and those pertaining to the associated external agencies.

We are, however limiting our discussions here to the 'Industrial Relations' issues pertaining to the employees and their representative forums, in the following paragraphs, for a general appreciation of the subject:

2.6.1 Personnel Policies

The starting point of '**Industrial Relation**' in any industrial enterprise is already incorporated in the **Personnel Policies** adopted by the enterprise management, particularly the policy frame-work formulated and adopted towards:

(i) Recruitment process and compensation packages;
(ii) Posting and deployment;
(iii) Promotion policy/career growth modalities.
(iv) Amenities and perks, welfare measures, social securities;
(v) Incentive schemes / reward schemes;
(vi) Training and skills development, re-training, re-deployment;
(vii) Retirement policy and terminal benefits, post retirement benefits;

(viii) Retrenchment policy, and terminal conditions;

(ix) Disciplinary matters; prescribed work-ethics.

(x) Work-place security and safety.

Employees, by and large, judge an employing organization and the management with respect to the above issues and form an attitude which have great influence and impacts on their behavioural pattern towards the industrial relations matters.

2.6.2 Participative Style of Management

Experiences gathered over the last four decades in some successful corporations in India have shown that employee participation in day to day operational affairs, which affect their working life and performance, have substantially improved the industrial relations, productivity and a sense of belonging / loyalty. Depending on as to whether the enterprise is a Public / Govt. undertaking or a Private Sector organization, the degree of employee participation in running the affairs of the enterprise may vary, but the positive impacts are visible and discernible, wherever this policy has been adopted with certain degree of commitment from the top management.

2.6.3 The Employee Participation Process, by and large, may get into practice through the following approaches:

(1) Recognizing Employee Representative forums like Trade Unions, Executive Associations, Supervisor—Associations, etc—within the frame work of relevant statutory laws and regulations and the company's declared policies.

(2) Providing necessary facilities and infrastructure to the above employee forums, to the extent feasible, with respect to the statutory provisions.

(3) Formation of a **Joint-Committee** at **the Apex Level** (particularly if the corporation is a multi-unit and/or multi location conglomerate, employing a large work force).

(4) Formation of a plant level **Plant Council** encompassing various major functional areas, again with declared objectives and terms of reference and with supporting organizational systems and procedures put in place.

(5) Formation of **Shop Councils** at various shop levels with declared objectives and terms of reference, and supporting procedures. **These shop councils are expected to function within the terms of reference of the Plant Council to avoid conflicting and divergent approach.**

2.6.4 Besides the above mentioned processes, certain other approaches, as mentioned below, circumstances demanding, can also be put into practice*

(i) Special Task groups may be formed to go into certain specific issues that may crop up and suggest remedial / corrective steps.

(ii) Task-Groups/Bi-partite Committees can be formed for Re-creative/ Cultural and Sports activities to improve/enrich quality of life, besides making positive impacts on attitudes towards work culture and team spirit.

(iii) Task-groups/Bi-partite Committees may be formed to monitor and sort out issues/suggest improvement measures for quality, safety, productivity as well as for employee welfare matters like medical, residential, schooling (of employees' children), canteen services etc.

****NOTE: Plant Council and Joint Committees as mentioned above may also deliberate on these issues and suggest remedial/ improvement measures, besides other operational matters like target fulfillment, productivity, quality, disciplinary matters, customer commitments, national and social commitments etc.***

2.6.5. The Participative Approaches mentioned above may help in achieving effective results by way of:

(i) Decision Making process quicker and implementable without much hitch.

(ii) Sanctity of Decisions: wide acceptability and respect amongst all section of employees.

(iii) **Participation based on equal opportunity** reduces grievances/ dissent.

(iv) Broad based Representation brings in added weightage and authority for implementation.

2.6.6 Extra-Ordinary Performance Recognition and Reward Schemes

Suitable performance recognition and reward schemes, formulated and put into practice, backed by appropriate systems and procedures, to make the same as transparent as possible. Opportunities to the talented employees should be provided to enable extra-ordinary performance in line with the set objectives of the enterprise.

- **Innovation and Creativity** must be recognized and contributions rewarded; the results achieved out of these may immensely benefits the organization—both in short term and long term perspectives.

- Adoption and successfully practicing a suitable Industrial Relations policy require certain degree of commitments on the part of owners/ management.

2.7 AWAY- CENTRES AND SUBSIDIARIES

In a large multi-unit industrial corporation, besides the directly administered units, there can be **'Away Centres'** and **'Subsidiaries'** operating more or less independently, but under the overall operational command and control of the main corporate body acting as a sort of a **'Holding Company'**. Such organizational set- up may be adopted to take the benefits of resource sharing, cost control, corporate tax advantages, employment rationalization, market-sharing etc.

2.8 EXTERNAL LINKAGES AND RELATIONSHIPS

Engineers working in industrial enterprises must also get familiarized with the processes of maintaining functional relationships with certain external agencies as mentioned in paragraph 1.4 in Chapter 1, but such dealings should be within the mandate given by the appropriate authority in the organization (seek guidance from the HOD as and when such occasion arise; **do not commit unilaterally beyond your mandate).**

"An Organization is what the people working there make of it."

—taken from a brochure of a Management Consultant Group

■ ■ ■

3

Mechanical Engineering and Design Aspects: Familiarization with Fundamental Approach

"Design is a strong capability since it allows continuous up-gradation of products and services".

—G. Raj Narayan, MD, Radel Group.

Young Mechanical Engineers who aspire to make a career in engineering goods manufacturing enterprises, should have, besides other related aspects, some knowledge on the fundamental approach to mechanical engineering and design processes that are essential for the conception of products/projects, product–development and configuration processes, and eventual product realization through appropriate technological processes on the shop floors. In the following paragraphs, we are attempting to present, briefly, relevant aspects, on the above topic for a general appreciation by young mechanical engineers:

REMAKS: After acquisition of academic qualifications in Mechanical Engineering fields, proficiency in applied engineering and design capabilities can only be acquired through practically doing such jobs over a period of time on projects or product-specific assignments. An appreciation of the following aspects is expected to accelerate the pace of acquisition of a reasonable level of knowledge in the field, when dove-tailed with the academic knowledge inputs already received.

3.1 MECHANICAL ENGINEERING AND DESIGN ACTIVITIES

Mechanical Engineering and Design activities in a composite industrial enterprise may be broadly grouped under the following sub-heads:

(i) Mechanical Project Design;

(ii) Mechanical Structural Design;
(iii) Mechanical Machine Design /Product Design;
(iv) Mechanical Services Systems Design;

In the following paragraphs, we are taking up the above topics one by one for the benefit of the young professionals:

3.1.1 Mechanical Project Design Activities

(A) Mechanical Project Design may be for

(i) A Turnkey Project,
(ii) An Individual Plant /Equipment/Machine,
(iii) Mechanical Services facilities,
(iv) Product Design (for regular production).,
(v) Process Design, (technological process elaboration, flow chart etc.).
(vi) R&D Projects for Product Development,
(vii) Reconditioning and Retrofitting of existing plant & machinery.

(B) **For all the above**, certain well established and standard design procedures can be followed and these procedures shall be broadly guided by:

(a) The kind of project and the desired end-results.
(b) Selection criteria for plants, machinery, equipments services etc. vis-à-vis the end-product and/ or services/output desired to be realized/ delivered.
(c) Kinematics and Linkage Mechanisms (**Kinematic-pairs**) involved in the Construction/configuration and the desired working principles.
(d) Standards and Codes, – both National and International as applicable.
(e) Procedures for design calculations, vetting procedures, and related work-testing if any required to establish desired product performance.
(f) Prescribed fits and tolerances as per the applicable standard/customer requirements.
(g) Quality Assurance procedures (Quality Plan) and prescribed/desired Accuracy level.
(h) Trial, testing and acceptance procedures.
(i) Finalization of project layout and technological flow paths, and preparation of layout drawings.
(j) Acceptance Norms vis-à-vis the customer requirements/preferences/ standards applicable..

(C) Project design must also take into consideration: ecological, environmental factors (pollution avoidance/minimization), better energy efficiency, energy savings and safe workability, economic maintainability, as important aspects.

(D) Young Engineers are expected to study and understand the approach, logic, relevance and purpose of each step in logical sequences,-by studying the documents/charts and by interacting with seniors, customers, may be also with the Consultants (if engaged) and any other agency(ies) involved; while doing so, they can always draw upon their academic theoretical knowledge acquired in academic courses **(although in practice, they may find that many of the activities being done do not exactly follow the theoretical formulae but on certain proven empirical formulae/ derived relationships, based on past experience and conventions, and the results coming out of dedicated R&D efforts undertaken).**

(E) Academic courses in Mechanical Engineering at the under-graduate level may NOT include much on project design aspects; therefore, young engineering professionals engaged in such project-work may have to heavily depend upon the experiences of the seniors, the prevailing working systems and the documentation on the subject available in the in the organization/department where they are attached to. The beginners will have to acquire knowledge and expertise through practical work-assignments, besides doing a lot of studies of relevant books, publications, reference documents etc,. and also through regular/frequent interaction with knowledgeable seniors/experts.

NOTE: (i) It will be useful to spend at least two-three hours a fortnight in the Organization's Technical Library (in case there is one): Technical Libraries are meant to help enhancing/up-grading knowledge that in turn help enhancing competence; hence career progression as well.

(ii) THE PROBLEM IS, THESE DAYS, most young qualified professionals after passing out and getting employed, STOP reading books/literature on relevant subjects thinking that they already have workable knowledge for their routine assignments; this attitude stifles urges for knowledge gathering for up-gradation of professional competence essential for career growth, particularly in the era of competitive career progression.

3.1.2 Mechanical Structural Designs (Using Steel/Alloy Steel, other Metallic Components)*

***NOTE: During the last four decades or so, many non-metallic* materials such as composites/synthetic fibrous compounds,** special varieties of glass, ceramic and silicon compounds, special grades of plastics etc have been developed as engineering materials including those for structural applications. **Please refer to Chapter 11 for more on Engineering Materials.**

(A) **Structural Engineering** as a subject and the associated design activities are, by and large, covered by a set of common procedures, shared by both, Civil and Mechanical Engineering branches. While, by a conventional division of works, the civil design engineer deals with the metallic (steel mainly) structures of buildings, bridges, scaffoldings, towers etc., the Mechanical design engineers may do such structural design for industrial plants and machinery such as boiler-house structures, gas plants, oil rig structures, structures for steel making plants, cement making plants, mining machinery, port and jetty structures, machine bodies, structures for sugar mills, flour mills, power plant machinery, machine tools etc. However, the basic principles of structural design for steel and other metallic structures for both civil and mechanical engineering applications will be the same/similar and these will have to take into consideration:

- The purpose/ end use,
- Loading magnitudes and loading patterns,
- All static and dynamic forces to which the structure shall be subjected to (moments, shearing, vibratory, seismic, storm- instigated impacts, gravitational etc.). Well- established/proven methods to be adopted for necessary analysis and calculations.

*(**Note: Appropriate software packages are now available for computerized configurations, structural analysis and design optimization).***

- Grades of steel or other ferrous (including alloy steels) and nonferrous metals to be used and their mechanical, metallurgical and chemical properties.
- Ambient Environmental conditions (variation in temperature and pressure, humidity, corrosiveness etc.)
- Recommended/prescribed Factor of Safety, Fits and Tolerances, applicable Standards. (National and International)
- Recommended/estimated safe working life.
- The kind of foundation and anchoring required.
- Besides basing the design calculations (for forces, stresses, strengths, sizes , configurations etc.) on theoretical relationships and formulae, certain empirical formulae derived from experimentation and past working experiences/performances, are also used, but one must be sure about the authenticity and respectability of the source of such empirical formulae/derived relationships before one can adopt the same.
- The stress should be on proven characteristics and applications.

(B) The advent of **CAE** (Computer Aided Engineering) and **CAD** (Computer Aided Design) have greatly enhanced and simplified the engineering and designing processes for which more and more advanced softwares are now being developed and made available; simply by changing the variable inputs, different variants as well as optimal solutions can be worked out within a short period of time. **Now a days, computer-literacy(applications) is essential for engineers and therefore, right from the beginning , fresh engineers are advised to acquire the same along with proficiency for practical aspects of computer applications.**

3.1.3 Mechanical Machine Design/Component Design

Mechanical machine design is rather a vast subject by itself and there can be endless lists of machines, plants, equipment, devices, gadgets etc. which may relate to the following functional/application areas:

(i) Material Handling Equipment

EOT-Cranes, Jib Cranes, tower cranes, cargo and passenger lifts, winches, transfer trolleys, lifting hooks/lifting beams, hydraulic lifting devices/scissor lifting devices, mining shovels, excavators, front-end loaders, Port Jetty based tower cranes, gantry cranes etc.

(ii) Transportation Equipment

Motor vehicles of all kinds, diesel locomotives/shunters, railway wagons/bogies, transport air-crafts, ships, motor boats, conveyor systems, MRTs (Mass Rapid Transport Systems) in urban/sub-urban areas- in and around big cities), coaches for under-ground (metro) railways, trailers etc.

(iii) Industrial Plants and Equipment of various kinds

For steel plants, power plants, engineering goods manufacturing plants, cement plants, plants for gas, pharmaceuticals/drugs, sugar, food processing, water supply, fluid flow, hydraulic and pneumatic devices, ship building yards, Railway-work shops, Naval work shops, aircrafts manufacturing plants, furnaces, ovens, foundary equipment, metal and non-metal extrusion plants, road building equipment, printing machinery, air compressors, gas compressors, pumps, mining equipment, material testing equipment, radiographic equipment, reprographic equipment, pollution control equipment/devices, heat exchangers, reaction chambers, scrubbers, coal washers, other ferrous and non-ferrous metal extraction Plants, pressure vessels, defence equipment (guns, rifles, rockets, missiles, battle tanks etc.), Nuclear Reactors and their auxiliaries; inspection and quality control devices/equipment etc., ………. **and the list can be endless.**

(iv) Machine Tools

Machines for Metal joining, metal cutting, metal working/forming, wood working, fibre/composite materials cutting, forming/working etc.

(v) Various types of tools, jigs and fixtures

Small tools, hand tools, machine tool accessories, job-holders, jigs, fixtures, dies, tool holders, metrological equipment /devices, measuring instruments.

3.2 THE FUNDAMENTAL APPROACH IN MECHANICAL DESIGN

Most of the above mentioned plants, machinery and equipment will predominantly comprise of the following category of mechanical components/ parts for which design works will have to be done by mechanical engineers:

(i) Static Components/Parts

Subjected to various **static forces**, loads and/or variable loads and stresses (mechanical and/or thermal or electromagnetic, hydraulic, pneumatic or a combination of some or all of the above).

(ii) Dynamic (Mobile) Components/Parts

Subjected to various dynamic forces and loads arising out of: reciprocal motions, rotary or vibratory motions, linear or non-linear, or curve-linear or, circular, or planetary motions or a combination of some or all of the above-with resultant mechanical, or thermal or hydraulic, pneumatic, or electro-magnetic stresses, or a combination of some or all of the above.

(NOTE: A reasonable knowledge on Application of Kinematics/Kinematic Pairs will be essential).

NOTE: Both static and dynamic parts will also be subjected to gravitational and frictional forces; also the dynamic parts, particularly the rotating ones, may additionally be subjected to windage forces (wind resistance/wind friction).

(iii) Miscellaneous Aspects of Mechanical Machine Design

Besides the above factors, the mechanical design of machinery and industrial plants will have to also take into account the prevailing working/ambient conditions: temperature, humidity, pressure, altitude, corrosiveness/other environmental impacts etc., as well as any other external forces and stresses that may exert influence on the working of the machine, equipment and plant.

The following are also to be taken care of:

(a) Ventilation, Cooling, Lubrication: These are also vital aspects which machine designers will have to take into consideration in the total system design:

- Assessing the requirements of proper ventilation in the work place, design and selection of ducting and fans with drives and controls; in corrosive environment, ducting and fans may have to have proper anti-corrosive lining with fibre glass/ plastic sheets, or ceramic materials, and may also have to have appropriately selected anti-corrosive coatings applied.

- Assessing the cooling requirements: cooling media, volume/ discharge rate, pressure, ambient temperature range etc.; and designing and selecting coolant and coolant system (pipe lines, pumps, drives and controls, valves, actuators, instrumentation, heat dissipators/heat exchangers etc.).
- Assessing lubrication requirements: rate of flow/discharge, volume, pressure, temperature etc.; and designing and selecting the lubrication system (piping, pumps, valves, drives, controls, instrumentation, selection of Lube oil grades, heat dissipators, coolers etc.)

(b) **Application of Standards**: For the design calculations and selection criteria for materials of construction, auxiliary support/sub-systems, must follow guidelines and procedures prescribed in National and International Standards and codes indicated therein. Relevant softwares may be used for CAD/CAE for design calculations, optimisation, sizing, configuration etc. A brief on application of standards in industry is given in **Chapter 14.**

(c) **Application of Kinematics and Linkage Mechanisms/Kinematic Pairs:** For Mechanical Design Engineers, particularly those working on machine/plant design, fundamentals and working concepts of Kinematics/Kinematic-pairs, linkage, link/lever mechanisms are essential, Although this field of specialization has a very vast scope, even then continuous upgradation of knowledge about advanced systems and improvements should be a way of life for the mechanical designers. It is expected that young mechanical engineers will have relevant theoretical instructions and practical exposures in respective academic technical institutions; nevertheless, it will always be useful to recapitulate on these very fundamental but essential aspects of machine design. For effecting various types of motions for performing different functions by a machine (or by its associated working systems), the following **Kinematic Linkage Mechanisms/Kinematic Pairs*** in various configurations are adopted/made use of in construction and configuration of the machine elements.

REMARK:** ***The listing given below is mainly for the purpose of general awareness; Those who are interested in depth knowledge about this subject, may read relevant books available in the market/

libraries. A few books are mentioned in Annexure-RB as reference reading materials.)

(i) Elements of Mechanisms:
  (a) Kinematic pairs (single-motion, two motion, three motion etc.)
  (b) Movable Joints (Two motion, Multiple motion etc.);
(ii) Single Lever Mechanisms;
(iii) Link Work Mechanisms;
(iv) Link Gear Mechanisms;
(v) Slider- Crank Mechanisms;
(vi) Lever-Cam Mechanisms;
(vii) Gear- Lever Mechanisms;
(viii) Lever- Ratchet Mechanisms;
(ix) Flexible-Link Lever Mechanisms;
(x) Elastic-Link Lever Mechanisms;
(x) Wedge-Lever Mechanisms;
(xi) Lever-Screw Mechanisms;
(xii) Simple- Gear Mechanisms;
(xiii) Lever- Gear Mechanisms;
(xiv) Pin- Gear Mechanisms;
(xv) Ratchet- Gear Mechanisms;
(xvi) Cam- Gear Mechanisms;
(xvii) Worm- Gear Mechanisms;
(xviii) Complex-gear Mechanisms;
(xix) Simple-Cam Mechanisms;
(xx) Cam-Lever Mechanisms;
(xxi) Cam-Gear Mechanisms;
(xxii) Cam-Ratchet Mechanisms;
(xxiii) Simple-Friction Mechanisms;
(xxiv) Complex Friction Mechanisms;
(xxv) Single Flexible-Link Mechanisms;
(xxvi) Complex- Flexible-Link Mechanisms;
(xxvii) Simple Hydraulic and Pneumatic Mechanisms;
(xxviii) Level – type Hydraulic and Pneumatic Mechanisms;
(xxix) Toothed-Hydraulic and Pneumatic Mechanisms;
(xxx) Elastic-Link Hydraulic and Pneumatic Mechanisms;
(xxxi) Complex- Hydraulic and Pneumatic Mechanisms;
(xxxii) Simple Electric Mechanisms;
(xxxiii) Lever-Type Electric Mechanisms;

(xxxiv) Toothed- Electric Mechanisms;

(xxxv) Complex-Electric Mechanisms;

(d) Fits, Tolerances, Limits: A brief on this topic is presented in Chapter 7.

3.3 MECHANICAL DESIGNS FOR DRIVES AND MECHANICAL POWER TRANSFER DEVICES

These are also very important aspects of mechanical designs of machines and plants—**specially machine tools.**

(1) Drives/Prime-movers

The drives are selected depending on the driving power (torque, pushing or pulling force, hammering force etc.) required and the methods of drives. Driving prime-movers are selected depending on the operational requirements (for linear, non-linear, curvilinear, rotary, reciprocal, planetary motions etc.) and also keeping in view the convenience of deployment. At present, the most convenient and energy-efficient prime mover is the **Electrical Motor** (Both AC and DC). As mentioned in **Chapter 10** under the head **"Prime-Movers"**, petrol/diesel/steam-engines, steam and gas turbines, hydro-turbine etc. can also be used as prime movers for drives, but none of them are as convenient as the electrical motors. However, when driving power required is enormous like those for large power generating sets, electrical motor may not be suitable for various limitations including availability of adequate electrical power and the large size of the motors. Design calculations carried out should enable selection of appropriate drive and driving power and hence the capacity of prime-movers. Special attention is also to be paid to design and use of energy-efficient drives for cost-effective operation. In fact this aspect is now becoming one of the important aspects in machine design and their operating cost. In addition, aspects relating to safe working and pollution minimization/ elimination (noise level, objectionable emissions, residual discharges etc.) are also to be considered seriously.

(i) Mechanical Power Transfer Systems:

For mechanical power plants, machine, equipments, devices etc., the most commonly used methods are :

- Line-shaft with flat-belts and pulleys.
- V-grooved pulleys and V-belts.
- Gear trains and couplings,
- Clutches and couplings (hydraulic, pneumatic, electro-magnetic, pressure plate friction clutches).
- Hydraulic and Pneumatic cylinders, pistons/rams.

- Through adoption of appropriate kinematics pairs and linkage mechanisms.

Any or a combination of a few of the above methods can be adopted depending on: the amount of power to be transferred, type of machine/ equipment deployed, the over-all convenience, economics of design and construction/configuration , maintainability, and operating costs.

3.4 MECHANICAL DESIGN OF MACHINE TOOLS

(a) This is a specialised branch of mechanical machine design and most of the aspects described above under paras 3.2 and 3.3 (Mechanical Machine Design/Component design) shall apply in the case of machine tools as well, specially the elements of kinematics and linkage mechanisms in particular. Therefore, mechanical design engineers dealing with the design of Machine Tools, besides being conversant with the aspects mentioned above under paras 3.1.3, 3.2 and 3.3; must also be conversant with the purpose and applications of various machine tools, the associated drives and controls, accuracy requirements, and the applicability of National and International standards for the same **(including those for Fits and Tolerances: Chapter 7).**

(b) The continuous advancement in the production technology in the last four decades or so, a host of new machining and manufacturing concepts have come into practice and these have, in turn, stimulated machine tool manufacturers to bring into market newer versatile and special purpose machine tools some of them for multiple operations. With the introduction of CNC machine tools in the last four decades, the versatility as well as operational accuracy, efficiency, productivity of machine tools have increased outstandingly. These have brought into effect further tightening of machine operating parameters requiring sophisticated drives and power transmission control systems and operating regimes, specially for the multi-axis operations.

(c) **Materials of Construction:** Generally for construction of machine tools, steel castings, grey-iron castings, steel/alloy-steel forgings, special grade steel structural materials are used, selectively for static and dynamic parts depending on duty-cycle and loading patterns, as well as considering ambient/prevailing working environment/ conditions. Selection of such metallic materials will require good level of knowledge on Metallurgical Characteristics and properties of the same. **For economic optimization on use of materials, application of Value Engineering Concepts will be important to achieve cost-effective /competitive design and configurations of parts/ components and assembly schemes.**

(d) **Lubrication and Coolants:** While the principle of machine-design as stated under paras 3.1.3, 3.2, and 3.3. above shall generally apply in the case of Machine-tool designs as well, special attention is to be given for **lubrication and coolant systems designs** for trouble-free operation. Where continuous lubrication is required, necessary lube oil pumping and pipeline systems with associated tanks/receptacles etc. will have to be incorporated in the total system; similar is the case with the coolant system.

However, care has to be taken that:

- Suitable arrangements are to be incorporated to retrieve used lube oil and coolant **(separately)** for possible recycling/disposal (this adds to economy in operation and cost saving).
- Proper grades of lube oil and coolant (with recommended additives if required) have to be specified by the designer of the machine/equipment.
- **Cyclic Lubrication**: For a large machine tool or mechanical plant, where a considerable volume of lubricant is required to be supplied to various locations/working areas, in order to economize on use of costly lubricants, a Sequential and/or Cyclic Lubrication system may be adopted with proper distributed drives and pipe lines with dedicated controls, instead of adopting a centralised lubrication system which not only requires higher energy consumptions but also higher volume of lubricants, and hence, chances of higher rate of losses as well and higher operating costs.

3.5 DESIGN OF TOOLS, JIGS AND FIXTURES TO BE USED ON/WITH MACHINE TOOLS

(a) For proper use of Machine Tools and other production equipment, specially those deployed for metal cutting, metal forming, metal working, essentially require tools, tool holders/collets/Mandrels, arbors, job-holding clamps, Jigs, Fixture etc. for carrying out the machining/forming operations on components/parts. There are standard tools, as well as there are special and/or non-standard tools also, depending on operational requirements. **Tool design is a specialisation by itself** and requires special skills and methods. Since the accuracy regimes and tolerances are very tight in case of toolings, and since these tools have to work under different loading and ambient conditions, design features for these require special considerations: the material of construction (grades of steel, forged or rolled stock etc.), heat-treatment processes, machining and grinding methods etc. Moreover, there are National and International Standards regarding the sizes, length, cutting angles, rake angles, accuracy, taper (such as

BT-50, 1S0-40 etc.), holding devices, adopters etc. which the designer must keep in view. The calculation of cutting, forming and working forces matching with the job requirements is also very important for tool design **(dedicated computer soft-ware are available and one has to be educated/trained on application).**

(b) (i) **For Press Tools/Dies** (to be used on power or hydraulic presses for forming or punching, trimming) as well as for the forging dies (to be used on forging hammers/machines), a different set of design considerations and calculations will be required to be done and material of construction selected accordingly as forces and stress acting on these are much higher.

(ii) **Tools and dies for die-castings, plastic mouldings** and for metal/non-metal extrusion processes are special purpose production aids, requiring specialised design and technological skills.

(c) For fresh engineers assigned to such design work must study and understand the methodics, regimes and basic principles of design applicable. Obviously, they have to draw support and guidance from senior experts, gather experience by getting involved in actual working, and even get practical tips from the technicians and operators who are already on the job for long years. (***False*** Vanity Prevents ***Useful Learning).***

3.6 MECHANICAL DESIGN OF FURNACES/OVENS

Furnaces are used for heating, stress relieving, melting, heat-treatment, baking, hardening, tempering and annealing etc. for ferrous and non-ferrous metals and also for certain non-metals as well, depending on the process requirements.

(a) Ovens are used for heating and baking, curing, thermosetting of predominantly non-metallic materials; Small ovens can be used for heating/heat treatment/drying/removal of grease/ oil, volatile substances from work-piece.

(b) Furnaces: depending on the temperature of operation and the heating medium used, the size and the purpose, the design features of Furnaces and Ovens and their construction features will differ to some extent, but the basic approach to design shall be more or less the same. The following working regimes of furnace operations will have to be considered :

(i) Kind of work-pieces/components, their sizes (dimensions) and weight.

(ii) The purpose: heating, melting, tempering, hardening, stress relieving, drying, baking, curing etc.
(iii) Heat load and temperature ranges.
(iv) Heating media (Electrical, furnace oil, coal gas, Natural gas, LPG etc.)
(v) Thermal insulation to be used;
(vi) Fixed hearth, or bogie hearth;
(vii) Continuous operating or intermittent operating.
(viii) Air blasting/Air supply arrangements.
(ix) Flue-gas discharge method.
(x) Fuel supply and fuel preheating and combustion methods.
(xi) Power supply for drives, controls, heating elements in case of electrical furnaces and ovens.

(c) The Furnaces will also have electrical engineering and instrumentation systems designs involved besides the mechanical design aspects.
(d) Ovens are generally smaller in size and may be used for both metals and non-metals; design considerations will be more or less like those for furnaces, but for a lower temperature range (100^0C to 500^0C normal ranges).
(d) Both furnaces and ovens will have controllers, indicating and recording instrumentation (small ovens may not have recording facility). Furnaces may have PID-Controllers with digital/analog instrumentation.
(e) Temperature and heat control sensing is done by thermocouples and/or **Pyrometric** devices connected to them.

NOTE: More about Furnaces and Ovens is given in Chapter 10.

3.7 MECHANICAL DESIGN OF BOILERS (FOR STEAM GENERATION)

(a) Boiler is basically a combination of the features of a furnace, a heat exchanger and a pressure vessel for fluids. Since steam generation is generally at super heat temperature and high pressure, boiler design will have to take into account all features of a furnace operation together with characteristics of a heat exchanger as well as the design features for the boiler drums as high pressure high temperature pressure vessels. Boilers will have supporting auxiliary system/services like water supply, fuel supply, storage and handling of fuels, ash and slag disposal, flue gas/exhaust discharge system (stacks), control systems, station power supply system etc.- all these sub-systems/auxiliaries will involve the specific working by mechanical, electrical, civil, electronics, and mechanical designers; therefore, supporting services of civil, electrical, electronics and instrumentation designers will also be essentially required.

(b) Mechanical design calculation methods will follow certain standard procedures depending on the desired parameters and operating conditions.

(c) Design and sizing /configurations of various pipe lines associated with boiler/auxiliaries will be an important aspect of the composite operating system.

3.8 DESIGN OF HEAT EXCHANGERS AND PRESSURE VESSELS

(i) Heat Exchangers

Heat Exchangers work on the principle of heat transfer from one medium to another inside a pressure vessel which can be a cooler or a steam generator or an economiser,—for utilization in plants like a thermal power station, petrochemical complexes, air and gas compressor plants etc.

The design calculation will have to take care of operating parameters (temperature, pressure etc.), nature of the fluids to be handled, material of construction, besides safety aspects.

(ii) Pressure Vessels

Pressure vessels may be for storage/in-process storage/reactor chambers, etc. where liquid or gas is stored, passed through, mixed at high pressure and sometimes at high temperature also. Mechanical design will have to take care of selection of materials of construction for the vessel, characteristics of the fluids, calculations for forces and stresses, operating regimes etc. matching with the parameters prescribed, besides safety aspects.

3.9 DESIGN OF MECHANICAL SERVICES FOR VENTILATION, AIR CONDITIONING SYSTEMS

(i) Ventilation

In many factories/plants, fumes, gases, vapours, dusts, smoke etc. get generated as a result of working processes followed. These cause unhealthy and uncomfortable work environments. To quickly evacuate such polluting and detrimental gaseous substances from the working area, adequate forced ventilation system with fans and ducting of proper sizes must be provided for. If required, adequate number of cyclones, scrubbers, and dust collectors are also installed. All these devices require design work and engineering calculations by experts for sizing and structural configuration.

Aspects which needed to be considered are: kind of such fumes, dust, gases etc. and their estimated volume/rate of discharge, size of ducting and material of ducting, lining of ducting if required, blasting/flow power required for displacement, and selection of corresponding fans with matching drives and controls etc. Engineers need to understand the approach/methods and logics of designing and design calculations, the relevant standards and codes to be used.

(ii) Air Conditioning

Whenever temperature and humidity are to be controlled, either below the ambient temperature, or above the ambient temperature (where ambient temperature is low below comfort level), air-conditioning units/plants are required to be installed either for industrial applications, or for human comforts, or for both. While for human comforts, window type AC units or Split AC Units can be used, the same may not be useful for industrial use as these window-type or split type do not control humidity below say 65-70% (RH) as well as the ambient temperature within the prescribed limits because of either larger volume of space or higher prevailing ambient temperature conditions, or both conditions prevailing. Therefore, for bigger halls/buildings and for certain industrial shops and enclosures, stores etc., we have to go in for Centralised AC units having higher capacity coolant gas compressors, ducting and chilling plants etc. These Centralized AC plants not only maintain the desired lower temperature but also controls the humidity within the desired RH(Relative Humidity) values (can be of the order of 35% to 60%) in the Main frame computer rooms, for industrial electronic control rooms, electronic telephone exchanges etc.). Besides the temperature (20°-22°C), the RH should not exceed 60% at any time as above this value of RH, life and performance of electronic elements will be adversely affected. In some CNC Machine-tool Controls, special compact insitu AC units are installed: these use special grade coolant gases to be highly effective in spite of small sizes. For some small storage room where certain special category heat and humidity sensitive materials with short shelf-life are stored, window type AC units may be used alongwith dehumidifiers to control R.H. within 50% or below, but in this arrangement, the load on the AC unit will increase and additional capacity may be required. The designers will have to consider all the ambient and environmental conditions PLUS the desired controlled conditions, and then design/select units accordingly.

Based on the heat load and RH calculations, designer can select from standard ranges of units marketed by various AC plant manufacturers. The young professionals are expected to understand the calculation methods, logics, and standards being followed in such exercises (heat loads, RH value, areas and volume of enclosures, variations permissible etc.).

NOTES: (i) In the last two decade or so, the technological processes for air conditioning and refrigeration have undergone certain changes which were forced by International ban/restrictions on the use of CFC gases (Chlorofluorocarbon: Trade named Freon-12, Freon-22 etc) as cooling/ chilling medium in the air-conditioning and refrigeration processes since CFC discharged in the atmosphere cause increased rate of ozone layer depletion in the upper atmospheric region, resulting in ozone devoid holes through which biologically harmful cosmic radiations pass and travel to earth surface thereby exposing people to cancerous skin diseases. The process for cooling/chilling now adopted are mostly based on vapour absorption technology .The readers may read books on modern air conditioning and refrigeration technologies for a better understanding of the subject (Please refer to the listing of books on the subject given at Annexure-RB). Designers of air conditioning and refrigeration systems are supposed to keep abreast with the latest technological developments in this field and create and maintain information/data-bank/documented records.

(ii) Please refer to Chapter-10 for more on Air Conditioning System.

3.10 MECHANICAL DESIGN OF HYDRAULIC AND PNEUMATIC SYSTEMS

In industrial set-ups, certain plants and machinery employ the power of hydraulic fluids and compressed air for effecting certain movements, transmission of power and for movement controls. Hydraulic Jacks, hydraulic cylinders, hydraulic pumps, hydro-static clutches compressed air pumps, compressed air operated hand tools, compressed air operated clutches, lifting and clamping devices, actuators (in fluid flow circuits) etc. find very common uses in industrial operating systems. Designer of such systems and equipment must have adequate knowledge of fluid mechanics, hydraulic/ flow characteristics of various fluids, purpose of applications, volume and pressure of fluid transfer, ambient and working temperature, and the physical and chemical characteristics of the fluid being used (there are a variety of hydraulic fluids/oils now being marketed by certain reputed manufacturers). Design calculations will have to take care of , besides fluid mechanics, strength of material aspects of the material of the construction of equipment and pipes used for hydraulic cylinders, pumps, hoses, control valves etc. Designer can also select certain standard designed components (available as bought out items) for the proposed set up.

Compressed Air Compressors of standard design are also available from certain reputed manufacturers, which the designers of the system can select as per needs. Designers are supposed to keep themselves up-to-date with the latest technological developments in the field and be adequately conversant with the National and International standards applicable to ensure quality, reliability and safe working regimes.

3.11 MECHANICAL DESIGN OF MATERIAL HANDLING SYSTEMS

Material handling is one of the important aspects of operations in an Industrial Project/Operating Unit.

Material handling activities in industry will mainly comprise of:

(i) Handling of construction materials;

(ii) Handling of raw materials, components and finished products on the shop floor;

(iii) Handling of materials at the customer project sites (during erection and commissioning of the same, in case such a provision is included in the contract).

3.11.1 Mobile Cranes/ Gantry Cranes/ Tower Cranes

(i) These handling equipment are generally employed outside the factory buildings for handling materials during construction, open yard storage, lifting of materials from or to railway wagons, truck-trailers etc. These equipment are available in standard sizes from certain reputed manufacturers. While design detailing are taken care of by the manufacturers in their organization, users may select them from the standard models and capacity ranges and can procure from the market. Drawing specifications for selection and procurement is a job to be undertaken by the project engineers in conjunction with the user department, plant maintenance department and purchase department. User and procurement agencies are supposed to collect necessary information and data from the prospective suppliers, manufacturers (technical brochures etc) since these collected details can help in firming up requirements and drawing up detailed specifications.

(ii) The designers of the manufacturing enterprise must be conversant with:

- Types of loads and loading patterns, magnitude of loads;
- Design and engineering procedures and practices relating to the product;
- Procedure for sizing and configuration based on customer requirement/market demand trends;
- Procedure for selection of materials of construction, selection of drives and controls;

- **National and International standards** relevant to the product including those for testing, quality control and safe working regimes.

3.11.2 EOT Cranes, Self-Standing Jib Cranes, Stacker cranes, Wall-mounted jib/cantilever cranes, suspension type cranes, lifting beams etc.

(i) These are normally used inside the factory blocks, but EOT cranes can also be mounted outside factory buildings on open to air gantry structures extend out of the factory block in a loading-unloading bay for convenience.

(ii) These cranes used in industrial units are normally procured in standard sizes from certain reputed suppliers and users need not undertake to do any design work. However, engineers engaged in the user industry/organisation have to draw up detailed specifications after assessing the applications, loading patterns and location of installation and based on this, procurement action is taken by the purchase department. The scheme of deployment of such cranes has to be decided keeping in view the shop-layout, technological flow of input materials, movement of components and loading-unloading at different points and movement of the finished products. However, special requirements of jib cranes and suspension type cranes, which are generally of low capacity (up to say 3 tonnes), can be met by in-house design (provided expert and confident designers are available in-house) and fabrication, procuring bought out items like hoists and its controls etc. from the market.

(iii) As regards the design and engineering activities in the manufacturing enterprise, the conditions stated at para-3.11.1 (ii) above are also applicable in this case.

3.11.3 Intra-Bay and Inter-Bay transfer trolleys

(i) These bogie type trolleys (or transfer cars) with flat platform at the top are deployed for handling and movement of materials across the bay where use of EOT cranes is not possible or not convenient. The load capacity can be from 2-3 Tonnes to around 100 Tonnes or more capacity. These trolleys can also be designed and fabricated in-house, or procured from outside agencies. Mechanical design engineers do the design calculations based on : the load carrying capacity desired, the size of the platform, height from the rail, the wheel size and number of wheels and axles required, gearboxes, materials of construction, drive/prime movers alongwith relevant controls to be used, ground conditions, rail size adoptable etc.

(ii) The conditions stated at para 3.11.1(ii) above also apply here as well.

3.11.4 Fork lifter, Frontend Loaders

(i) Fork lifters upto 4 tonnes capacity are available in battery operated models; beyond 4 tonne, they are driven and operated by Diesel Engines. Front-end loaders are higher capacity loading-unloading mobile machines driven and operated by Diesel engines and may have hydraulic cylinder/ram operated mechanisms as well. User industries do not design them as these are available in standard design and capacity from reputed manufacturers who have their own design engineers for this purpose. However, it may be useful to know the principles and methods for design of such machinery.

(ii) The conditions stated at para 3.11.1. (ii) above also apply here as well.

3.11.5 Conveyor Belt Systems

(i) This is a widely used system of material handling (loading-unloading and bulk transfer). The belts are made of corrugated thick rubber sheets and moves on a series of rollers supported on suitable structures. There can be metallic (steel, alloy-steel) grating type conveyers also moved by chain and sprocket system and driven by Electrical Motor(s), specially for hot-zone working or for hot materials. These can also be are driven by diesel engines or steam engines. These can be procured from certain reputed manufactures who can design, manufacture and install such systems as per requirements. They will have their own design engineers to carry out the design and prepare drawings for production on shop floors. However, rollers and drive systems may be procured as bought out items from other firms specialising on manufacture and supply of such items. User organisation need not have their own design engineer, but engineers drawing up specifications for such system must be conversant with the critical factors (which have a bearing on the design, capacity and reliability etc.) while undertaking procurement activities for these.

(ii) The conditions stated at para 3.11.1.1(ii) above also apply here as well.

3.12 MECHANICAL DESIGN OF TRANSPORTATION SYSTEM

Transportation of materials and personnel,—both are necessary for smooth operation in an industrial enterprise. While vehicles such as cars, pickups, buses or even trucks (petrol or diesel engine driven) are deployed for movement of personnel, trucks /tailors, dumpers, conveyors, transfer trolleys etc. are mainly used for transportation of materials, both within the premises or outside. This is done either by owned vehicles or by hired vehicles. For long distance transportation, besides deploying trucks, materials can also be conveniently transported by railway wagons or even by air-cargo transport planes. Since user organisation does not design and manufacture such vehicles or railway wagons (unless the same is in their product ranges), these are procured from reputed manufacturers, or taken on hire from transportation companies or from Railways (as the case may be). However, when the user organisation goes in for procurement, its engineers must know as to how the specifications are to be drawn up (which requires some knowledge of design and operating parameters of these vehicles/equipment).

However, the design aspects will have to take into consideration the loading and speed regimes besides structural design and compatibility and all other related systems and parameters of operation, besides safety and reliability aspects.

3.13 DESIGN OF FOUNDRY EQUIPMENT (USED FOR MANUFACTURE OF CASTINGS AND FORGINGS)

(i) Foundries require certain special types of machinery and equipment besides the normal facilities like handling equipment (EOT Cranes, transfer trolleys), heating/reheating furnaces, melting furnaces, machine shop equipment (like machine tools etc.), welding and gauging equipment etc., Cupolas, Electric Arc furnaces, Holding furnaces, various types of ladles, sand preparation plant, sand reclamation plant, moulding machines, pattern making machines, fettling equipment, flame heating equipment, Vacuum Degassing Units(VDU), Vacuum Oxygen Decarbonisation Unit (VOD), Vacuum Degassing/ Casting Facilities (VD/VCF), Radiography equipment, forging presses/ hammers with matching manipulators, special hooks on EOT cranes (for handling hot metal ladles, hot billets/blooms and for handling hot-job pieces during forging process), pneumatic vibratory feeders, pneumatic grinders, water quenching/mist quenching equipment, oil quenching tanks, high capacity air blowers, electro-slag remelting equipment (ESR) etc.

(ii) The special purpose foundry equipment mentioned above are procured from reputed manufacturers who have their design experts. User organisations need trained engineers and foundry technologists to successfully and gainfully utilise these equipment for productive outputs. However, engineers dealing with foundry technology must have adequate knowledge of metallurgical processes and have the capability to draw up detailed specifications of various equipment for successful execution of procurement, installation and commissioning (total dependence on foundry equipment suppliers may not be prudent as they may take undue advantage of lack of knowledge and skill on the part of engineers/technologists of the customer/procuring organisation.**(This is, of course true for all engineering and technological enterprises).Association** of Metallurgical Engineers will be imperative in such tasks.

3.14 MECHANICAL DESIGNS FOR POWER STATION EQUIPMENT

(NOTE: Readers may refer to Chapter 12 on 'Energy/Power Supply Systems' in this book wherein the basic concepts of such setups have been dealt with.).

(i) The Mechanical Design aspects in Power Plants will have to deal with:
- Thermal Power Stations(steam based):Boiler House equipment and their auxiliaries; steam generators/HRSG, Steam turbine and auxiliaries.
- Hydraulic Turbines and their auxiliaries.
- Gas Turbines and their auxiliaries.
- Mechanical design aspects of Electrical Generators and their auxiliaries.
- Mechanical design aspects of Nuclear Reactors and auxiliary equipment (for Nuclear Power Stations).
- Mechanical design aspects of Diesel Power Plants;
- Mechanical design of associated auxiliary services such as fuel handling and supply systems,. ash and slag handling and disposal system, cooling water and DM water supply systems, flue gas discharge system, ESP system, station substation structures and such others.

(ii) Design procedures, will generally follow the same approach as mentioned above for machine and equipment designs (paras-3.2, 3.3, 3.4 above). **However, besides the above,** environmental engineering aspects will require special attention. For Nuclear Reactors, special skills and knowledge are required for successful execution and operation of the plants, keeping in view safety of operation and statutory requirements of ecological factors. Only the trained and skilled experts can do such jobs.

(iii) Even for Electrical Generator and other electrical machines specially for the rotating machines, of body construction, supports, bearings, cooling and ventilation system, lubrication system etc. are very critical and have to take care of the static and dynamic forces and stresses, strength of materials, constructional feasibility etc.

(iv) While the **user organizations** may not carry out such design activities and leave it to the manufacturers, but engineers engaged by the user organization must also know the basic design features before they can draw up the purchase and the installation specifications, performance and acceptance test regimes etc. in order to execute the project and operate the plants smoothly and negotiate with/monitor the activities of the manufacturer vis-à-vis the specified requirements including quality and reliability aspects. It is always advantageous for engineers of user organizations to understand the basic design aspects and parameters that the manufacturer takes into consideration for designing the equipment.

(v) In case of **Nuclear Power Stations**, leaving aside Reactor House equipment, other power station equipment are by and large similar to those in the conventional thermal power stations, and the approach to designs will be similar.

(vi) **Coming to Reactor House Equipment**, the Reactor Vessels, the coolant pumping and piping system, loading-unloading system for the fuel rods, controls etc., require very specialised designing/engineering capability and technological skills; also because of presence of radio-active materials, they have to take into consideration, not only the equipment/plant safety, but also of biological and environmental safety aspects. The technology and design details are rather closely guarded strategic matters and not readily shared by the concerned engineering and manufacturing organizations. However, apart from the engineers and the technologists from the manufacturing/supplying organizations, the engineers of the user organization are given only those technological information which may be just enough for evaluation of bids, purchase decision, for installation, commissioning, and for day to day operations (including upkeep and maintenance of the plant). In any case, engineers of user organization has to undergo certain special training with the manufacturing organization for acquiring certain level of expertise and skills for successful operation and maintenance of the plants.

(vii) Steam generation takes place in a specially designed Heat Exchanger called **'HRSG' (=Heat Recovery Steam Generator),** normally installed in a separate building adjacent to the reactor building. HRSG is an **Unfired Boiler** which utilizes the heat generated in the Nuclear Fission Reactor, heat being transferred from the reactor through the heated coolant circulating in the primary piping system which transfer the heat to the water circulating in the secondary piping system where the water gets converted into super heated steam; this steam is used to drive steam turbine that in turn rotate the coupled turbo-generator to generate electricity.

3.15 MECHANICAL DESIGN OF PRIME MOVERS

The readers are advised to read the topic on 'Prime Movers' in **Chapter-10** this book and the relevant paragraphs on 'Mechanical Designs of Drives' in para-3.3(1) above to get an idea of broad out-lines on these machines. In case the organisation itself is the manufacturer of such prime movers, then its design engineers are expected to be conversant with various factors mentioned above under 'Mechanical Machine Design and Component Design' (para-3..3 above). These are continuously developing fields and designers must keep themselves continually updated on various upcoming technological developments and operating aspects, be aware of market/customer requirements: lack of this aspect as well as outdated designs may force the organization out of business.

Even in the case of the organization being an user one, its engineers must be conversant broadly with the design aspects and the parameters on which the manufacturers are basing their design calculations, otherwise they will find it very difficult to:

- Draw detailed tender specifications for purchase;
- Carry out techno-commercial evaluation of offers received as well as doing negotiations before finalization of order.
- Check quality and reliability aspects.
- Decide acceptance norms for taking over after erection and commissioning.
- Successfully operate and maintain.

"The ability to identify problems and devise imaginative responses to them is crucial to good performance."

—Loren Gary.
(Harvard Management Update)

■ ■ ■

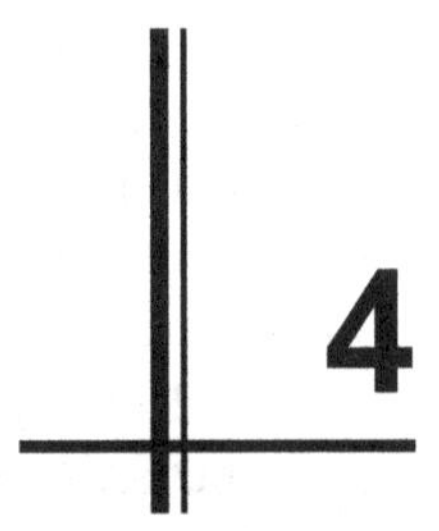

Mechanical Work-Shop Practices for Manufacturing and Also for Plant Maintenance and Repair Services

"Engineers are educated and trained to be inquisitive, analytical and meticulous about whatever they are assigned to do, of course through legitimate means, with a disciplined scientific mindset."

—Author's view.

REMARK: The scope of discussion in this Chapter relates to activities that take place in an engineering goods manufacturing industrial enterprise. The coverage of topics mainly aims at RE-CAPITULATION of what has been learnt during academic studies and dove-tailed into the practical industrial working processes. Many of these workshop practices are also applicable for repair jobs specially for capital repair. Readers are advised to read Text Book on the subject (Please see Annexure-RB).

4.1. WORKSHOP PRACTICES

(i) In any industrial workshop/manufacturing unit, certain standard shop processes are, by and large, universally being practised. With continuous development and introduction of new technologies, and occasionally new products, the shop practices undergo certain forced as well as innovative changes. However, many age-old well-established shop practices in the traditional ways still continue with attendant advantages. These shop practices in a manufacturing unit are closely linked with prescribed/desired productivity, product quality, functional reliability, aesthetics and the overall techno-commercial competitiveness of the end products.

Mechanical Engineers working in such shops must continuously keep themselves updated with latest practices and ensure quality and economical operations by judicious choice of methods and relevant tools; they have to take upon themselves the onerous task of identifying the comparative benefits of old and new practices, educate and encourage workers to adopt the best methods for better productivity and quality for meeting customer requirements successfully in the face of the competitive markets situations.

(ii) Engineers also have to see to it that wastages and rejections are minimal and that the workers are conforming to the prescribed systems and methods, besides ensuring clean and safe working environments in the work-places; untidy and unorganized conditions on the shop floors cause a de-motivating environment and sometimes also demoralize the workforce. (Please refer to the **5S work culture presented in Chapter 13**).

4.2.

The Entire Gamut of the Manufacturing Processes for production of engineering goods such as structures, machinery/equipment/components, spare parts etc., the processes and activities which convert raw materials and semi finished items into finished products, are referred to as the **'Production Technology'** which may be grouped into **three broad classes**: **Shaping, Joining, and Finishing**. These three main shop processing activity-groups can be further split into sub groups as under:

(A) **Shaping:** { Machining, Forming }

(B) **Joining:** { Fastening, Welding }

(C) **Finishing:** Assembly, Testing, Surface treatment, Packing and Shipping.

NOTE: Engineering goods manufacturing units predominantly make use of metals and their alloys which again comprise of ferrous and non-ferrous metals in various grades.

4.3 METAL WORKING PROCESSES

In the metal related production technology, to prepare and process the metals into the desired shapes and sizes, in conformity with the prescribed metallurgical and environmental conditions; in order to give effect to the above mentioned three **broad activity-groups** as mentioned in para-4.2 above, two sets of processes are followed; **these are:**

(A) **Hot Working Processes comprising of:**

(i) Rolling, (ii) Forging, (iii) Extrusion,
(iv) Piercing, (v) Drawing or cupping, (vi) Spinning,
(vii) Welding, (viii) Die-casting, (ix) Plastic Moulding.

(B) **Cold-Working Processes comprising of:**

(i) Cold Rolling, cold pressing,

(ii) All kinds of metal cutting, machining, metal forming/metal working under ambient temperature conditions,

(iii) Swagging (a kind of cold rolling),

(iv) Shearing (v) Drawing.

Diagram No. 4.1 symbolically represents the various manufacturing technologies adopted in engineering industry.

Manufacturing Technologies (Conceptual Representation)

Cuttig Processes

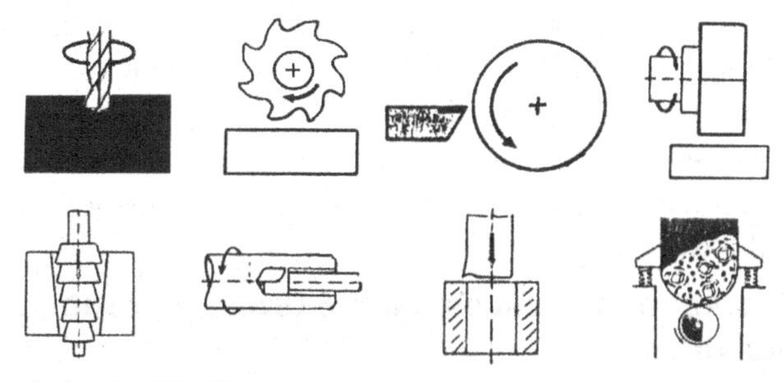

Non-Cuttig Processes

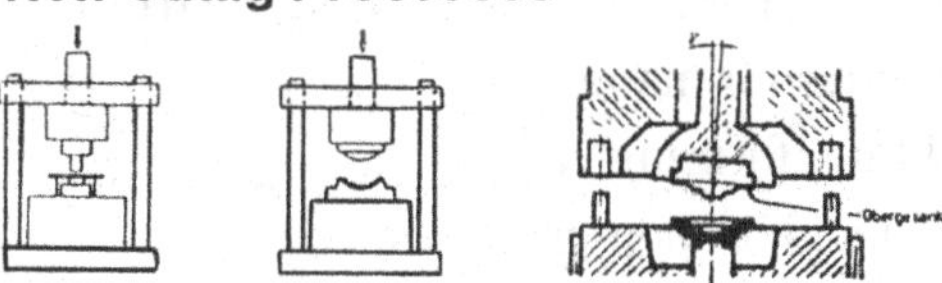

Heat treatment

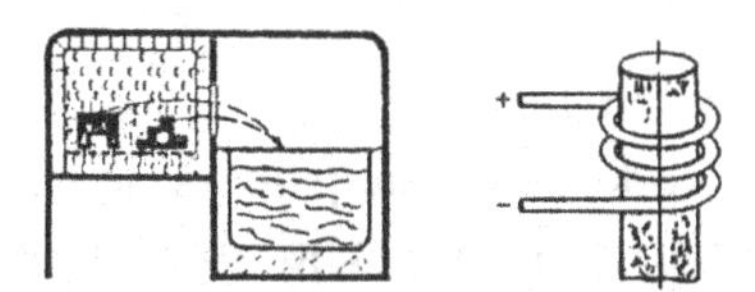

Other production techniques

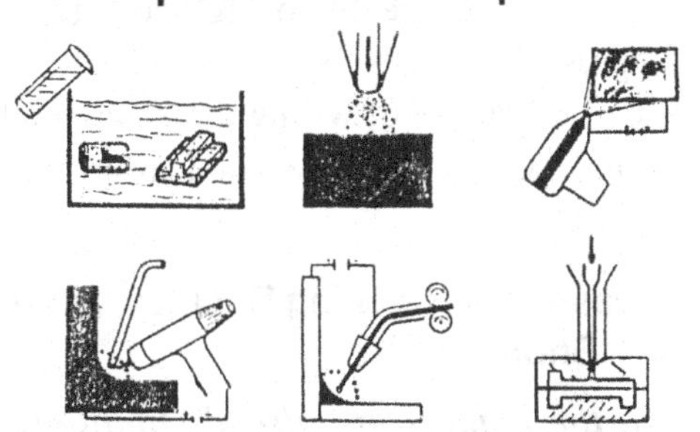

Source: Taken from the Brochure of a Machine Tool Supplier

Diagram No. 4.1

4.4 WORKSHOP PRACTICES

The workshop practices and the manufacturing activities for metal based industries mainly deal with the processes of metal cutting, metal forming/working, metal shaping, assembly, finishing/surface finishing, testing and other down-steam processes. In the succeeding paragraphs, we shall discuss various aspects relating to these processes, besides other related matters.

It is expected that the young mechanical engineers have already undergone certain level of acquaintances in production technology, workshop and machine shop practices in the respective technical institutes, and that they are, by and large, conversant with the methods, tools, tackles, machines and equipment etc. being used there. However, it may be useful to have a re-capitulation of these in the following paragraphs for the benefit of the young engineers to help in successfully starting an Industrial Career with confidence. It is also to be kept in mind that besides the knowledge of technology, it is very important to know the proper use of various machines, tools, jigs &fixtures, and supporting/associated equipment, devices etc., as these are very critical for successful operations in the workshop.

4.4.1 The Workshop Practices can be broadly grouped under the following activities:

(1) Use of Small Tools and Measuring/checking Instruments.

(2) Preparatory activities for manufacturing.

(3) Fabrication, welding, forming etc.

(4) Machining and Machine shop practices.

(5) Fitting and assembly.

(6) Shop Testing.

(7) Surface Treatment, Packing and dispatch.

Besides the regular production works on the shop floors, some of the above shop practices are also applicable for repair and major overhauling works on existing plant, machinery, structures (capital/fixed assets in particular) to keep them in working order for productive usage. Shop engineers must also see to it that workers use proper and prescribed tools to get desired results and to avoid rejection/reworking because of use/abuse of wrong tools and methods.

We shall now briefly discuss each of the above listed group of shop activities as mentioned below:

(1) Use Of Small Tools, Measuring/Checking Devices, Instruments/ Shop Accessories In Work Shops

NOTE: It is presumed that the readers are already familiarized with the fundamentals of Workshop Practices in their academic inputs. Also, they read Text Books on Workshop Practices for recapitulation to supplement knowledge;

(1-A). **Small Tools and Devices for Work-shop Practices:** The following small tools, which are mostly hand-held/hand operated/portable, are widely used for various initial/smaller fitting and assembly jobs; some of them, of course, are also used for final fitting and assembly of medium and large components as well:

- **Different types of Hand-held Hammers** (depending on usage): Ball-peen, Straight peen, cross-peen, Claw hammer, Soft-faced hammer or Mallet.
- **Hand Punches:** -Drift punch, Pin punch, Prick punch, Centre punch, and Automatic centre punch.
- **Scribers:** to draw lines on surface.
- **Compass:** to draw circles/semi-circles on surfaces;
- **Screw Drivers:** Light duty/heavy duty screw driver, Phillips type screw driver, helical ratchet screw driver with automatic quick return, double ended offset screw driver.
- **Pliers:** Slip point, Needle nose, Linemen's side cutting Pliers, (some pliers are electrically insulated with plastic or PVC covers on the handle), diagonal pliers.
- **Wrenches:** Single ended, double ended, closed end wrenches, twelve point box wrench, monkey wrench, lever-jaw wrench, pin hook spanner wrench, adjustable hook spanner wrench, adjustable hook spanner wrench, adjustable pin-face wrench, T-socket wrench, offset socket wrench or L-end wrench, strap wrench, pipe wrench, chain pipe wrench, Pipe cutter, hydraulic and torque wrenches.
- **Work Holding Clamps:** Tool maker's clamp, C- clamp, V-blocks with clamps, Toolmaker's hand-vice, drill-vice, bench-vice, pipe-vice, chain pipe vice etc.
- **Scrapers and Snips:** Tinsmiths snips, bearing scraper, three-cornered scraper, burr removing scraper.
- **Tool Boxes:** Machinist's Tool Box, Fitter's Tool box, Electrician's Tool Box.
- **Chisels and Chippers:** Flat cold Chisel, Cape chisel, round nose chisel, diamond-point chisel, electric chiselling and chipping hammers, hot chisel.
- **Hacksaws, Jig Saws :** Hand operated, with matching cutting angle must be known to the workers to avoid infructuous efforts and rejections.
- **Files:** Mill file, Flat file, Pillar file, Square file, Round file, Three Square file, Half-Round file, Knife file, Triangular file, Swiss pattern crossing file, Swiss pattern needle files, Vixen flat file, Cabinet file, Flat rasp file , Flat lead float file ; File cleaning steel brushes, Lutz file handle, filing machine, diamond files.

- **Soldering Irons :** Electric soldering irons/guns, blow torch for gas/oil flame metal soldering.
- **Hand Operated Taping and Threading Handles and associated items:**

  To be used with proper knowledge of internal and external threading process, thread angles and thread pitches.
- **Tap extractor** (when tap gets broken and stuck up inside the thread holes)
- **Ezy-outs** for above (with reverse spiral movement).
- **Threading dies** for external threads.
- **Screw pitch gauge/thread micrometers, ring thread gauges**.
- For small components, thread can be checked on optical profile projectors or on tool-maker's microscopes for correct profile, angle and pitch of tools.
- Thread can be checked on comparator, thread lead gauge, pitch diameter thread gauge, centre gauge.
- Threads can be cut on lathes with proper thread-cutting attachments.
- Threads can be formed by using thread rolling machines (which uses standard thread rolling dies), or by using a thread whirling device.
- Threads (high accuracy) can be ground on thread grinding machine to obtain high finish for accurate angles and pitches.
- **Ring gauges and plug gauges** are also used to check correct size of threading.

(1-B) Measuring Tools and Instruments, shop accessories, used in workshops

(i) Measuring dimensions and determining dimensional and machining accuracies of parts/components is very important as any deviation beyond the permissible limits may lead to rejection/rework and therefore, result in loss of time and money. Shop engineers must ensure that they are fully aware of the measurement requirements and procedures, as well as arrange to provide proper measuring tools/instruments/devices to the workers/inspection agencies. They have also to ensure that workers and supervisors are properly trained to handle and use them, and that they are also conscious of the importance of using them as and when required. Without this work-culture and motivation, the whole manufacturing process may get jeopardized and the credibility of the shop engineers may suffer badly.

(ii) **The following Measuring Tools/Instruments/Devices are commonly used in any work-shop :**

- **Scales** (for linear, angular, circular measurements):-
  Triangular box-wood scale, standard steel rule, Hook rule, Narrow hook rule, Calliper rule, Rule depth gauge, combination gauge set, Box-wood folding rule.
- **Dividers, Compass, Tramel Compass.**
- **Callipers, Verniers, Micrometers:** Outside callipers, inside callipers, hermaphrodite callipers, micro-meter callipers/vernier callipers, screw thread micrometer, depth micrometer, indicating micrometer, tube micro- meter, inside micrometer.
- **Vernier Height Gauge:** Vernier Height measuring Dial Gauge, Off-Set scribers, Gear-tooth vernier, Vernier bevel protactor, Height Masters.
- **Level Gauges:** Spirit levels, straight edge beams, optical alignment telescope, optical projectors, strain gauges, laser alignment devices.
- **Hydraulic/Pneumatic Nut-tightener.**
- **Jacks** (Screw jacks, Hydraulic jacks),
- Augurs, Portable hand drilling machine, Portable electric drilling machine, bearing removing clamps.
- Steel rope slings, eyebolts, lifting beams, scissors lift tongs, pulleys, pneumatic holding and manipulating devices.
- **Gauges and Gauge Blocks:*** Ring gauges, Plug gauges, snap gauges, caliper gauges, dial indicator type gauge, indicating hole gauge, boroscopes, feeler/thickness gauges, perthometer (for surface finish measurement), indicating thickness gauge, indicating depth gauge, radius gauges, amplifying comparator gauges, precision gauge blocks (standard sets), Surface plates, Angle gauge blocks (standard sets), Surface plates, Angle gauge blocks (standard set).
  ***NOTE: Now a days,** electronic **digital display devices are available for convenient and reliable usage. CMM (Co-ordinate Measuring Mechanism with CNC System) are also used in large machine shops.**

(2) Preparatory activities for manufacturing (engineering products/ components etc):

(2-A) Before Starting/Launching actual manufacturing activities on the shop floor, it is essential to complete certain preparatory work so as to avoid/minimise hold-ups and bottlenecks in the progress of work. The **preparatory work** shall include:

(i) Formal Work-Order which serves as the authorization from the Management to launch/start manufacturing since this in turn requires commitment and utilization of Resources and incurring of expenditure/costs.

(ii) Receipt of manufacturing drawings and careful study of the same to get adequately conversant with the details.

(iii) Availability of the specified and relevant materials on the shop floor/shop stores matching with production schedules and in conformity with the bill of materials foreseen in the manufacturing drawings and related process documents.

(iv) Availability of the required toolings, jigs and fixtures, shop accessories.

(v) Availability of adequate number of trained workers/operators for deployment at the work-place.

(vi) Operational readiness of machine tools and other production/ process equipment.

(vii) **Quality Plans** and Acceptance norms, in documented/chart form, specifying parameters of acceptance criteria for the finished items.

(viii) Technological process documentation, route cards indicating norm-hours, material introduction schedules and the job movement schedules and routes to be followed.

(ix) Manufacturing schedules for various component and sub-components etc. (issued by **PPC group**).

(2-B) Pre-Launch Checks: Engineers posted in manufacturing shops must ensure that the above requirements are complied with well in time to avoid delays. Continuous stress should be laid on proper planning and control/monitoring. In case there are delays on the above inputs from any of the concerned department/agencies, the matter has to be followed up vigorously at various levels. **The engineers assigned with production jobs must also take particular care of the following:**

(i) Manufacturing drawings and technological process documents be checked well in time for any mistakes/mismatch, missing information, lack of clarity to avoid delays.

(ii) To check and ensure that all documents received are clear and up-to-date and relate to the relevant work-order and component numbers.

(iii) Materials being drawn from stores/sub-stores are of correct grades, size and category and in adequate quantity, but avoid drawing excess material (wastage, cost implications).

(iv) To check the control-chart schedules (issued by the **PPC** department) for various activities relating to the component and carry out updation on a day-to-day or week-to-week basis (as per the need of the system in vogue).

(v) To check availability, adequacy, and relevancy of the supplied tools and fixtures (lined up for use).

(vi) To ensure that **Quality Plan**/instructions on quality and prescribed tolerance limits are adhered to.

(vii) To be sure that workers working on various machines **clearly Know** as to what task to perform and how and in what sequence.

(viii) To ensure that all documentations and records are maintained and preserved till the prescribed periods of preservation to enable cross-reference, future investigative reference, traceability, resolution of customer complaints, and the like.

(ix) To ensure that **inter-shop** and **intra-shop** inter-action takes place regularly for information sharing and mutual tie up of inter-dependent activities/schedules (to avoid confusion, buck-passing etc.).

(x) To promptly tackle any other exigency that may occur to ensure adherence to schedules.

(xi) To have **intra-group** periodical meetings for a concerted approach and action for fulfilling the assigned tasks, in line with delivery commitments made to the customers.

(xii) **To avoid crisis situation developing at any point of time**, take prompt corrective action.

NOTE: To take adequate care of the issues mentioned at paras 4.4.1(1) and 4.4.1(2) above are not only essential for smooth operation of the unit, but are also very important in case the unit is already an ISO-9001 or TQM Certified enterprise, or is aspiring to get such certification in near future.

(3) Fabrication, Forming and Welding Processes

Since engineering industrial products (machinery, equipment etc) are predominantly made out of steel /alloy steel items (plates, flats, bars, rolled stocks, angles, channels, I-beams, billets, blooms, etc.), grey iron and steel castings, steel forgings etc., to make the desired components/parts, a series of fabrication activities are required to be carried out to give the component/parts the designed shape and size prior to assembly/machining of the same (some fabricated items may not require any machining and can directly get assembled into the final products/component with some fitting and finishing works done on these).

However, for some industrial products, components etc., certain non-ferrous metals like Aluminum, Copper, Brass, Bronze are also used and these also need to undergo necessary manufacturing process like the ferrous metals, with of course certain variations in the processing in conformity with their chemical and physical/mechanical/metallurgical properties.

(3-A) The Fabrication and Forming, by and large, comprise of the following processes:

(i) Cutting of blanks, rolled stocks, structural steel items, plates etc. by using:

- Hacksaws (reciprocating type)
- Steel cutting band saws/cold cutting circular saws etc.
- Oxy-acetylene flame cutting machines using suitable machine-held torches, PUG cutting machines, NC/CNC-flame cutting machines etc.
- Shearing machines, cropping machines such as hand shearing, heavier shearing machines using power (crank)/hydraulic presses depending on size and thickness of plates, bars, etc.

(ii) Forming of steel items by using- power (crank)presses, hydraulic presses presses (using dies); also using 3-roll/4-roll Plate bending machines.

(iii) Pipe bending machine, bar/rod bending machines.

(iv) Hammering for thin sheets, bars etc;

(v) Hammering for forgings : using Pneumatic hammers or Hydraulic presses , depending on type and size of forgings.

(vi) Fitting works and assembly as per drawings/process sheets.

(3-B) **Welding for joining fabricated items, rolled and forged items:** Welding of steel/ferrous/non-ferrous materials/pieces together for obtaining desired shapes and sizes.

(3-C) Out of all the above processes, welding of metals (ferrous, non-ferrous) assumes the most important method in fabrication. Modern welding methods have developed into different sub-branches depending on the materials and the technological process to be adopted, keeping in view the metallurgical properties (of raw materials, finished components),safety, reliability and quality desired. Before going into further details about various welding processes in vogue, the engineers engaged in the shop floor for fabrication works must take care of the following preparatory checks:

(3-D) **Pre-Fabrication Checks:** Before taking up fabrication works in the shop floor:

(i) Check the correctness/relevance of the **drawings and process** sheets available/supplied to the shop(ensure availability of the up- to-date versions).

(ii) Availability and correctness of grades and sizes of materials being drawn from stores/stocks for fabrication (select material as per the specified tailoring charts as far as possible to avoid wastages by way of higher level of scrap generation); **ensure**

selection of appropriate quality and grades of electrodes to be used for welding.

(iii) Use of proven/prescribed methods of welding process and availability of correct/proper welding machine, tools/tackles; clarity on limits prescribed/deviations permissible.

NOTE: It is also very important to know the techniques of retrieval/ rehabilitation/rectification works without rejecting the work-piece when deviation occurred.)

(iv) Get conversant with the prescribed stage-wise quality checks and ensure compliance.

(v) After-welding/fabrication, further processes to be followed (like heating and cooling for stress relieving, sand/shot blasting, machining, fitting, assembly, hydraulic testing etc.)

(vi) Availability of trained personnel for carrying out stage-wise inspection and quality control.

(vii) Agency to whom components/parts to be handed-over after completion of fabrication process.

(viii) Availability of skilled workers (fabricators, welders, press operator, gas cutting and saw machine operators, sheet metal workers etc.).

(3-E) **Checks to be done during Execution of Fabrication Works**

(i) Ensure that prescribed procedures are being followed and prescribed machines are being used to carry out the prescribed process.

(ii) Drawings are being properly studied and correctly interpreted/ understood.

(iii) Quality checks during the in-process stages are carried out against prescribed norms.

(iv) Production programmes/control charts are being followed and progress records maintained.

(v) Deviation, if any, allowed by the authorised agency, be recorded properly (for future reference).

(vi) Shop up-keep and safety aspects taken care of **(Follow 5S system: 13.2)**.

(vii) Timely inflow of material and tools, and proper grades of gases, be ensured.

(4) Welding Sciences and Metallurgy

In industrial manufacturing process, welding is one of the most commonly adopted method for durable joining of metal together, although there exist welding processes for non-metals (like plastic, fibre material, epoxy sheets etc.) as well. **We are presently limiting our discussions on the welding of steel structural components, machine components, other ferrous materials and**

components, as these constitute a very high percentage of elements on which welding processes are applied to.

Over the last four decades or so, the welding technology has diversified into several special methods, besides the traditional ones, to meet the ever increasing demands for higher productivity, quality and reliability of welded joints/components specially for construction of highly stressed steel structures and super-critical machinery parts etc. which are subjected to high degree of pressure, temperature, static and dynamic (including vibrations) stresses. **For the benefit of the young engineering professionals, it may be useful to briefly re-capitulate on the following topics:**

(4-A) **Welding Sciences:** Basic Principles of welding relates to:

- Joining of metals by heat-fusion.
- Joining of non-metals (plastic, fibre-glass etc.) by heat fusion.
- Joining of metals and non-metals under cold pressure (heatless welding: such as by ultrasound welding, by laser beams, by applying pressing forces).
- Joining metals and non-metals by methods that do not employ fastening devices.

(4-B) **Welding Metallurgy:**

- Weldability of metals (and non-metals also in some non-metallic application),
- Fusion heating welding deposit is a casting (the filler material from the electrode gets deposited as a casting fused with the surrounding body under heat –treatment process.)
- During welding by fusion heating, the temperature rises to about 800°C to 1100°C and at that temperature metal partially gets oxidized into oxides; this is detrimental to the strength of weldments.
- During fusion welding by heat, heated metal will dissolve gases, mainly O_2 , N_2 and H_2, water vapour from the surrounding air and discharge CO_2, CO gases as well as water vapours; even very small amounts of these ambient elements may ruin the quality of the metal deposit vis-à-vis, the metal surrounding the deposit. *Under the fusion welding heat, metal gets both oxidized and* embrittled, thus loosing the characteristic of ductility which is important for shaping/ forming metal bodies/components by bending/twisting/rolling etc., thus the metallic performance of the welded structures get degraded/ deteriorate.
- In the Fusion Heat Welding, the metal of electrode and that of the body to be welded can differ in metallurgical composition, and the resultant weldment may have a metallurgically inferior composition thus create a new degraded metal which may not meet the requirements of strength

and ductility for the welded structure; therefore, case has to be taken for adoption of appropriate welding process.

(4-C) **Welding Technology:** Welding Technology for engineering and industrial applications can be classified into the following four general types:

(i) Gas Welding;
(ii) Electric Resistance Welding (ERW);
(iii) Electric Arc Welding (EAW);
(iv) Special Methods of Welding (including Robotic Welding);

The types at (ii) and (iii) above are based on electrical power although many of the 'Special Methods' (iv above) also use electricity for the source of welding power.

(4C-1) **Gas Welding**

(i) Gas Welding utilises the energy released during the combustion of a fuel gas mixed with air or oxygen. For energy-efficient welding at high temperature, air/ O_2 mixed with other combustible gases like Acetylene (C_2H_2), Hydrogen(H_2), Propane (C_3H_8), Natural Gas(which is predominantly Methane or CH_4) or Butane (C_4H_{10})—is used through appropriate gas welding devices. Most commonly used gas-combination is Oxy-Acetylene for gas welding process. The rate of deposit of weld material in gas welding is slow and quantity-wise small; **gas welding is therefore, normally used for welding thin metal sheets, small diameter rods, small size bars, non-ferrous metals like brass, bronze, copper etc. in small sizes of sheets, rods etc.**

- **Gas welding is not suitable for larger and thicker/heavier metal pieces and structures.**

(ii) Gas welding is not performed on metals such as titanium, molybdenum etc. as these metals get metallurgically deteriorated due to presence of carbon and hydrogen in the fuel gas mixture.

(iii) Gas welding has also the disadvantage of heat not confined to the narrow spot of welding, resulting in substantial loss of heat and extra consumption of fuel gas.

(iv) Gas welding has a very good advantage in the fact that it requires very simple and minimal equipment and thereby making it handy and easy to use.

(v) Gas welding process generate much lower temperature than the Electric Arc welding and therefore, it is suitable for metals having lower melting points such as lead, tin, copper.

(vi) **Gas welding serves excellently** for such work as brazing, soldering, general repair and maintenance works, autobody works, plumbing of small piping, thin metal sheets, etc.

(4C-2) Electric Resistance Welding (ERW): In case of Electric Resistance Welding, the heat of welding is generated by flow of electric current in the secondary low voltage circuit of the AC system (using welding transformer) where the low resistance of the circuit plays an important role, together with the pressure and heat at the spot where welding electrodes press the two or more pieces of metal and the same gets fused by melting. This is also called **'Spot Welding'**. This method of welding is usually used for sheet metals (or thin metal pieces) welding; besides fusing the metal pieces together, external metal pieces like nuts, washers etc. can also be fused to the metal pieces. Two other versions of the same welding process, called the **'Seam Welding'** and **'Projection Welding'** are also widely used for sheet metal welding depending on the size and the configuration of the metal pieces to be Welded. Electric Resistance Welding (ERW) process is widely used for the manufacture of steel pipes and conduits from steel sheets but with a continuous stream of spot weldments at joining sheets. Another ERW process called the **'Butt Welding'** is applied to rods and pipes for end to end joining by self-fusion. ERW obviously, cannot be used for heavy plates or structural steel components.

(4C-3) Electric Arc Welding (EAW)

(i) **This is the most commonly used welding method** for light, medium and heavy welding requirements. This method uses electric arc for generating heat for welding. The arc is created across the electrode and the work-piece by appropriate AC or DC electric power supply system at a certain low voltage (40 to 60 volts). **The welding arc is virtually a sustained short circuiting phenomenon** used for generating intense heat which melts the electrodes and deposits weldments in the form of a casting on the body of the work-piece and this casting gets fused with the work-piece body, thus joining or filling gaps for structural requirements.

- **AC Arc Welding**: AC power is obtained from a transformer welding set with low voltage but high current at the secondary winding which gets connected to the electrode.
- **DC Arc Welding:** DC power is obtained from:

  Motor Generator Set.

  Trans-Rectifier Set.

(ii) AC welding using transformer set is more energy efficient and easier to use but there are inherent limitations for use in pressure welding with higher current level, may create spatters on the weldments, thus effecting the quality of weldments.

(iii) DC welding generally gives smooth weldments and is preferred for welding flat, vertical and over-head positions. **For non-ferrous metals, DC welding is more reliable and gives better quality of welded joints.**

(4-D) Sub-Categories of Arc Welding

(i) **Submerged Arc Welding:** In the process of Arc welding in the open air, the quality of welding may suffer because of hot weldments getting partially oxidized. To improve quality of welding, some method that is very commonly used is 'Submerged Arc Welding' where welding spot is covered by welding flux and the arcing and fusing takes place underneath the flux-cover, thereby eliminating (near to 100%) the chances of oxidation of hot weldments. In this case, welding electrode in wire form is continuously fed from spools to the submerged arc spot which also continuously gets submerged by falling flux dust from a chute: the entire system is mechanised by a suitable arrangement for semi-automatic or automatic operations and is employed when heavy metal deposit rates are required.

(ii) **CO_2 Shielded Arc Welding:** In this method, the Arc welding process takes place under an atmosphere of CO_2 gas, thereby eliminating/substantially reducing the chances of oxidation and thus maintaining higher quality of weldment. Here also, welding electrode wire is fed to the electric arc along with feeding of flux powder and CO_2 gas blown at pressure over the welding spot to prevent/minimise oxidation.

(4-E) **Special Welding Processes:** Certain metals (particularly alloy steels) require special welding process to achieve the desired degree of quality and finish of weldments. Sometimes after welding, machining of the welded joints may not be possible or desirable and therefore, finish of welded portion/joints must be good enough to meet the quality requirements. Some of the special welding processes that are now widely used in engineering product manufacture are mentioned below:

- TIG Welding:-Tungsten Inert Gas Welding;
- MIG/MAG Welding:- Metal Inert Gas/Metal Argon Gas Welding;
- Plasma Torch Welding;
- **Robotic Welding:** The above welding processes are a combination of gas and electric welding, since they use electric arc carried

through a stream of gas of special combination. Robotic Welding is widely used in fabrication and assembly of vehicular car bodies.

(i) **TIG Welding Process** is used for thin-gauge special alloy steels and non-ferrous metals with Argon or Helium as the inert shielding gas. Electrodes used in TIG welding are Tungsten-Thorium alloy and/or, Tungsten-Zirconium alloy. TIG welding may use ACHF or DC (DCSP or DCRP) power supply system. The electrode wire feeding and flux feeding are by semi-automatic devices (wires are fed from spools). >>**(ACHF=AC High Frequency; DCSP = Direct Current Same Polarity; DCRP = Direct Current Reverse Polarity)**

(ii) **MIG/MAG- Welding process** is also partly similar to TIG process, but is used mostly for non-ferrous metals like Aluminium, Titanium, Zirconium, Magnesium, Beryllium etc. with Helium and Argon as the shielding gases. MIG/MAG welding is a combination of electric short-circuiting and spray-transfer of metal in inert gas shielded coverage. Pure CO_2 gas is also used for MIG-weld shielding. MIG welding uses DC (DCRP) power supply for better quality of welds.

(iii) **Plasma Torch Welding:** This method is basically an extension of the TIG process and uses gas at a very high temperature to reach the plasma state when the gas becomes ionised and electrically conducting. The gas can be compressed air, CO_2 and Argon-Hydrogen or Nitrogen-Hydrogen combination depending on the material to be welded or cut. Plasma torch welding method is used for non-ferrous thin metal pieces which should get fused together **in a confined space very quickly under the plasma torch and thus avoid distortion and oxidation.**

(iv) **Special-Purpose Welding:** Besides the above, there are a number of special purpose welding methods, to name a few:-**Electron-Beam Welding, Laser Welding, Ultrasonic Welding, Robotic Welding etc.**

- **Robotic Welding process:** This is a process comprising mainly of spot welding, TIG and MIG/MAG welding set-ups with automatic/semi-automatic welding by the use of Robotic devices with **CNC system of remote controls**.

(4-F) Beside the technological aspects discussed above, the shop engineers must take care of the following:

(i) Ensure selection and use of proper welding process as prescribed for the component.

(ii) Use of proper grade and diameter of electrodes. Care should be taken to make identification of grades to avoid mixing up and wrong use.

(iii) Proper preparation of the metal to be welded before welding process starts: cleaning, sand blasting, edge preparation, grooving, sizing.

(iv) Proper up-keep and readiness of power supply sources of required voltage and KVA ratings.

(v) Availability of shop accessories, tools, tackles, safety shields/goggles, hand gloves, manipulators etc.

(vi) Proper working of ventilation system for quick evacuation of fumes, dust, smoke, etc.

(vii) Maintaining proper Arc Length during welding; failure to do so will adversely effect welding quality besides slow progress and delay.

(viii) Proper arrangements for protecting welding (trailing) cables from mechanical damages.

(ix) Selection of required grades of welding fluxes and proper storage of the same with correct marking to avoid mixing-up and deterioration in storage.

(x) Preheating/drying of flux coated electrode rods/flux powder (to drive away moisture/dissolved gases).

(xi) Proper earthing- point connections.

(xii) Measures for fire prevention and fire-fighting.

(4-G) Some other relevant points about Welding Processes which the WELDERS must be made aware of to understand the significance of:

(i) Welding Techniques: Arc welder must be familiar with the following aspects:

- The Polarity, arc-length, welding current, open circuit voltage, arc-voltages, type and size of electrodes, edge preparation, work piece setting, welding implements, welding machines etc.

(ii) Welded Joints: Lap-joints, Butt joints, Tee-joints, Edge-joints, Corner joints etc.

(iii) Edge Preparation: Single V, Double V, Single U, Double U, Single bevel, Double bevel, Single J and Double J.

(iv) Defects in Welding: Cracks, blowholes, slag inclusion, crater, undercut, spattering, excessive penetration, excessive reinforcement and overlap.

4.5 THERMAL CUTTING OF METALS

(A) In the fabrication process, sometimes large scale cutting of metals (steel plates, steel structural elements, rolled stocks, billets, bars, grey iron and steel castings, forged items) **is required to be carried out** prior to welding, machining, fitting and assembly etc. These cutting processes can employ either the machining process or **the thermal cutting processes** by using Oxy-fuel-gas flame (mostly Oxy-Acetylene) cutting, or plasma-torch cutting or arc-cutting methods, depending on the type and size of metals to be cut. **Thermal Cutting of metals** is not only expensive, but also slow, consumes large quantity of oxygen and fuel gas. However, the thermal cutting process is far more versatile and offers the following advantages :

(i) Thermal cutting process can be employed to cut metal pieces of great thickness, much beyond the capacity of shearing machines; cutting very thick metal by cold saws/band saws is very time consuming, besides there are other physical limitations on the machining capability; in fact for thermal cutting, there is no practical limit to the thickness of metal that can be flame cut.

(ii) Any shape can be cut by thermal cutting method, and almost in any geometrical line or curve depending on accuracy of profile required. (Tailoring chart used as required.)

(iii) In case of big and thick steel-plates, rolled bars, etc., the machine cutting may not be feasible because size limitations and over-all economy of operations **(capital cost of such extra heavy machines like heavy shear, plano-miller, planner, cold circular saws etc. will be very high, besides the high operating costs).**

(B) Apart from the most commonly used Oxy-Acetylene flame cutting, other methods employed for thermal cutting, depending on the metal category (specially thick special alloys, thick non-ferrous metals), are:-

- Powder cutting (employing flow of iron powder along with the flow of a fuel gas)
- Electric Arc Cutting (mainly for Steel plates).
- Plasma-Arc Cutting (Specially for thin non-ferrous plates for enabling quick cutting in a narrow cutting zone).
- Laser Beam Cutting- Using laser beams of high intensity to cut thin metal or non-metal sheets; this process gives very good cut-edge finish and no further machining is normally required.
- **Wire cut-EDM machines** for high accuracy and high finish cutting **(Note: More on this is dealt with in Chapter 5).**

(C) Nowadays, besides employing manual gas cutting equipment, **semi-automatic and automatic NC/CNC flame cutting machines** are also being widely used where substantial cutting on a regular basis is to be carried out. These have distinct advantage of accuracy of cut profiles, and higher productivity.

(D) **Shop Engineers need to take care of the following for trouble-free operations:**

(i) To select the proper method of thermal cutting, taking into consideration the type of metal to be cut and shapes to be obtained.

(ii) To ensure use of recommended type of fuel gas-mix and their adequate stock and supply; stock identification system must be in place to avoid wrong use.

(iii) Proper cutting torches and tools, proper fixtures/supports and their upkeep.

(iv) Safety aspects (accidents, fire hazards, electric shocks etc)

(v) Properly trained operators.

(vi) Use of tailoring charts etc. to reduce the level of scrap generation.

4.6 MACHINING PROCESSES FOR METAL CUTTING, METAL FORMING/WORKING IN ENGINEERING INDUSTRIES

These topics are being discussed in details in **Chapter 5**, since the subjects of Machine Shop Practices and Machining Technologies require rather elaborated treatment for familiarization and productive practices.

4.7 PORTABLE HAND TOOLS AND AUXILIARY WORKSHOP DEVICES

"Be faithful in small things because it is in them that your success lies".

—Mother Teresa.

Portable hand tools of various kinds and sizes are used in workshops for different purposes. **In fact, these small items of equipment are in great demand in different workshops in industrial complexes for their versatility and ease of operations leading to reduction in worker fatigue**. At the same time, these are a **most abused category** of devices on the shop floor, as there is a tendency on the part of the users/operators to take their performance for granted, often overlooking their capacity/working limitations and subjecting them to overload, even misuse. These result in frequent break-downs/damages, needing repair or replacement. In most cases, there is no deliberate effort for their regular preventive maintenance and therefore, they are mostly subjected to only break-down maintenance/repair, or in extreme cases, discarding them altogether for getting new ones.

With the above work culture prevailing in most workshops, we are discussing, in the following paragraphs their trouble-shooting and repair needs:

(NOTE: The readers may please read topic on '5S- System' and 'Jishu-Hozen' discussed in Chapter 13).

These hand tools and shop devices mostly come in the following operating categories based on their power source:

(i) Electrically operated;
(ii) Pneumatically operated (compressed air operated);
(iii) Hydraulically operated;
(iv) Engine driven Portable Tools/Devices

4.7.1 Electrically Operated Hand Tools

(i) Commonly used in industrial workshops are :

- Electric hand drills (upto 50 mm)/core drills;
- Electric hand grinders (straight grinders, angular grinders; can use stick/cylindrical type or disc type grinding wheels)
- Electrical Polishing and Buffing machines;
- Electric Hammer Drills;
- Electric Stone/Metal Sheet cutters;
- Electric Rotary Hammer;
- Electric Die Grinder;
- Electric Belt Sander;
- Electric Finishing Sander;
- Electric Air Blowers;
- Electric Nibblers;
- Electric Jig Saw Machine;
- Electric Circular Saw;
- Electric Chain Saw;
- Electric Cordless Drills;
- Electric Power Planers;
- Electric Portable Router;
- Electric Hand Beveller;
- Electric Saw Blade Sharpener;
- Electric Riveting Machine etc.

There can be many more.

(ii) Most of the above mentioned electrically operated portable hand tools /devices are powered by Universal Electrical Motors of capacity ranging from 200 watts to 2500 watts; weights ranging from 0.4 Kgs to 12-14 Kgs. Universal Motors (AC Commutator Motors which can work on both AC and DC power supplies; now a days mainly from Single Phase AC mains of 220/230 volts, 50/60Hz).* The speed ranges from 150 rpm to 16000/18000 rpm depending upon the type of the motor.

*(iii) Some of the above mentioned tools/devices such as Sanders, buffing machines, high speed grinders, polishers, blowers have a working speed from 3000 rpm to 18000 rpm depending upon the duty cycle required. To achieve such high speeds, special AC power supply installation having a system frequency of about 400 Hz is also installed in the concerned shop for these special purpose high speed tools.

NOTES: (a) For this purpose, specially designed medium/high frequency AC motor-generator sets are available in the market (brushless versions are also available).

(b) While selecting such portable electrically operated tools, care should be taken for operator safety and preference should be for low voltage power supply system (12 V to 48 Volts or around).

(iv) A few small devices like Cordless drilling machine operate on special purpose heavy duty rechargeable dry cells (4 to 7 AHs).

4.7.2 Pneumatic Hand Tools

These portable hand tools are powered by compressed air (normally 6 to 8 kg/ cm^2 pressure) supplied to the equipment by means of appropriate sizes and quality/grade of hoses, connected to shop's compressed air distribution mains(pipe lines running near the work centres).

The commonly used pneumatic hand tools used in industrial workshops are:

Pneumatic hammers, pneumatic hammer-cum concrete cutters, pneumatic chippers, pneumatic nut tighteners, pneumatic grinders, pneumatic Die-grinders, pneumatic hand polishing machine, pneumatic hand held buffing machines, pneumatic nailing machines, pneumatic hammer-cum riveting machines and the like.

4.7.3 Hydraulically Operated Portable Tools and Devices

A few commonly used ones in industry are :

(i) Hydraulic jacks of different capacity (ranging from few kgs to 50 tonnes or so; jacks above 20-Tonne are rarely used in Industry except for platform type vehicular/wagon weigh bridges the use of which are rare in engineering industry.

(ii) Hydraulic Lifting devices (mobile)/Scissor Lifts;
(iii) Hydraulic crimping tools;
(iv) Hydraulic torque wrenches;
(v) Hydraulic clamping devices; (for holding work pieces on machine tools; for jigs and fixtures).\
(vi) Hydraulic hand held gripping/pressing devices.

While some of the above devices are operated on Hydraulic power (pressure) generated by manually operated (lever operated) integral hydraulic pumping and cylinder piston system (like jacks, crimping tools, lifting device), many others draw their hydraulic power from portable hydraulic pumps (driven by either an electric motor or by small petrol/diesel engines) through appropriately connected hoses and hose couplings and matching outlet points.

4.7.4 Engine Driven Portable Tools

These devices are of limited use in logging/timber cutting and in construction projects where electric power supply is not readily available. Commonly used ones are:

(i) Chain saw/ Timber saw portable petrol engine driven.
(ii) *Engine driven rock cutters/driller/breakers;
(iii) *Engine driven blowers/portable compressors;
(iv) *Engine driven concrete cutters.
(*Mostly portable Diesel/Petrol Engines).

4.8 FITTING AND ASSEMBLY PROCESS; FITS, TOLERANCES AND LIMITS, ENGINEERING MEASUREMENT AND DIMENSIONAL METROLOGY, APPLICATION OF RELEVANT STANDARDS

These topics have been dealt with in **Chapter 7** and **Chapter 13**.

"There is no future in any job.
The future lies in the man who holds the job."

—George Crane

■ ■ ■

5

Machining Technologies and Machine Shop Practices (Metal Cutting/Stock Removal, Metal Forming/Working, Wood-working)

"Man-Machine compatibility by way of demonstration of technological knowledge, capability and application is essential for productive utilization of given Facilities and Assets for remunerative outcomes."

—Author's view

REMARK: ***The topics dealt with in the following paragraphs are basically meant for re-capitulation of knowledge inputs received during the academic courses, but the details sought to be presented here seek to emphasize on conceptual clarity for practical applicability; this is expected to enhance/ sharpen the subject knowledge of the young mechanical engineers who aspire to make a career and acquire higher proficiency in their working in the Engineering Industries.***

Those who are interested to know more about the subject, they are advised to read books on production technology (related to engineering industries) and Mechanical Engineer's Handbook besides interacting with knowledgeable experts and undergoing practical training; the topics dealt with in this book may help in providing certain practice-oriented knowledge inputs to the young engineers who are starting their career in engineering industrial enterprises.

5.0 MACHINING TECHNOLOGIES

In the **engineering goods manufacturing shops,** a variety of machine shop practices are adopted for machining of parts, components, jigs, fixtures, supports, work-holders, tool-holders, spare parts etc. For this purpose, a large variety of **machine tools and equipment/devices** are available in the market, and more and more with technologically advanced features are coming up every now and then.

With the development and entry of **CNC Machine Tools** in the last four decades, the whole concepts of machining has undergone revolutionary changes-, specially for better accuracy/quality, higher productivity and operating flexibility.

The CNC Machining Centres are designed and configured to carry out multiple machining operations even on one setting of the job; also the use of ATC (Automatic Tool Changer) on some CNC machines have increased their productivity and versatility; thus reducing the machining cycle-time substantially.

With the introduction and retrofitting of DROs (Digital Read-Out systems) even on conventional machine-tools, the accuracy level and quality of outputs could be improved considerably. Thus, the whole culture of machining processes and machine shop practices are undergoing sea-changes. The process of incorporating technologically advanced features will continue to improve productivity and quality/reliability, besides bringing economic benefits in the shop operations.

5.1 METAL CUTTING PROCESSES

There are a wide variety of basic and traditional machining processes being employed in a modern work shops.

5.1.1 Commonly Practised Machining Processes in the Manufacturing work-shops (which may include those for Capital Repair jobs/making Replacement Spares) are listed under:

(i) Turning,
(ii) Chucking,
(iii) Shaping,
(iv) Shearing, Punching and Trimming,
(v) Planing,
(vi) Facing/Face Milling,
(vii) Milling Operations,
(viii) Copy Milling/Tracer Controlled Milling,
(ix) Jig Boring, Profile Projection,
(x) Boring,
(xi) Grinding,
(xii) Broaching,
(xiii) Sawing,
(xiv) Drilling,
(xv) Taping,
(xvi) Threading/Screw Cutting,

(xvii) Slotting,
(xviii) Honing,
(xix) Nibbling and Routing,
(xx) Gear Hobbing,
(xxi) Gear Shaping
(xxii) Gear Shaving,
(xxiii) Lapping,
(xxiv) Super Finishing,
(xxv) Polishing and Buffing,
(xxvi) Abrasive Cutting,
(xxvii) Electro Erosion Machining,
(xxviii) Hydro-Jet Machining,
(xxix) Machining Centres.

- Besides the above, certain special machining processes are also adopted for specialised jobs; to mention a few well known ones: Laser Cutting, Trepanning, Electro-Chemical Machining, Rifling.

NOTES:

(i) For the benefit of the young mechanical engineers, a set of pictures of a few types of Machine Tools are presented at the end of this Chapter.

(ii) For the same type of machines, different manufacturers may adopt different design and configuration philosophies and try to incorporate certain special features to impress upon prospective buyers/users about additional benefits/advantages.

(iii) Mechanical Engineers engaged in Production Technology and Machine Shop practices must develop a habit of keeping themselves conversant and updated about new technological development trends in machine tools and their applications. There are a number of periodical publications on Machine Tools; besides, reputed manufacturers supply technical brochures of their products on request. An information data bank be created in the department since it is a convenient way to keep abreast with the latest developments in the field.

5.1.2 While the **general purpose machine tools** are used for machining a few pieces of jobs in one setting, CNC machine tools are normally employed for batch production; the semi-automatic or fully automatic machines are used for large scale series production in one setting.

NOTE: For more details about the NC/CNC machine tools application etc, please see Chapter-8 wherein Selection Criteria for machine tools and the cutting tools used on them have been described.

5.1.3 Types of Machining Operations

Depending on the purpose/end-use and the shape/configurations to be obtained, any or a combination of many of the above machining processes(para-5.1.1 above) are employed for rough and/or finish machining of components/parts. Let us now examine, in the following paragraphs, the applications of various machining processes, together with types of machine-tools/equipment that are being used for the same:

(I) Turning

(i) Turning is a very commonly adopted method for jobs requiring circular machining on outer surface, end faces, inside bores or hollow bodies. Machine tool used for this purpose may be any of these depending on the job-configuration to be obtained:
Centre Lathes, Turret Lathes, Facing Lathes, Screw cutting Lathes, Bench Lathes, Tool Room Lathes, Single Spindle Automatic Lathes, Multi-spindle Automatic Lathes, Contour Turning (copying) Lathe, Relieving Lathe.
With certain accessories and special attachments fitted on certain types of lathes, operations other than turning, such as boring, drilling, circular grinding, circular grooving, face-milling/grooving etc. can also be carried out by the use of relevant cutting tools and fixtures. Circular profile shaping can also be done on a suitable lathe machine with special tooling and fixtures.

(ii) All the **Lathe Machines** mentioned above are horizontal spindle type, but there are vertical axis lathes which are also called '**Vertical Lathes/Borers**', as these machines can do, besides outer and inner turning operations, other operations like milling (face milling, circular grooving, large diameter boring and even grinding of surfaces by a grinding attachment fixed to the tool post in certain medium and large sized machines.

(iii) The sizes of **Lathe Machines** can widely vary from bench types of less than metre length for very small/tiny jobs, to **large and medium lathes** (depending on the component sizes) in the ranges as under:

CD = 500 mm	SW = 250 mm
to	to
CD = 12000/15000 mm;	SW = 3000/4000 mm;

where, CD = Centre to Centre Distance;
SW = Swing Over Bed;
Bed lengths from 1500 mm to 36000/40000 mm.

(iv) Tools used on a Lathe Machine are predominantly single point tools of different working angles (setting angle, centering angle, rake angle, working-end cutting-edge angle etc.). While doing horizontal drilling or boring along the machine axis, common arbor type drilling tools or boring/milling(single point) tool, or even reamers can also be used with suitable tool holders fitted on the tool post or on the tail stock.

(NOTE: Readers may refer to para-2. in Chapter-8 for a brief on selection and application of cutting tools.)

(v) Depending on the type of metal and the finish required, the tools with required geometry and working angles are selected and set on the tool holder/tool post. The recommended standard-charts are available for this purpose (**consult Tool Engineer's Hand Books**); the required information can also be had from reputed standard tools manufacturers.

(vi) Chip-breaking is an important requirement during machining process to ward off interference/blocking and chattering affecting quality of finish. Chip breaking is therefore, now being incorporated in the cutting tool, either by making a small step or groove into the face of the tool bit, or by a separate piece of properly shaped metal piece fastened to the tool, or, to the tool holders to cause chips to break into short pieces or curls and get cast off automatically to the chip tray or chip conveyor placed down under the machine.

(vii) In single point tools, to economise on costs, the body (i.e. shank portion) an be made of ordinary steel, while the tool bits are made from high speed carbon steel, carbide steel, cast alloys or even from hard non-metallic materials; the tool bit can be fixed on the shank-tip by soldering, brazing, welding, or by suitably screwing. In this system, the tool bits can be re-sharpened and reused repeatedly before replacing them at a stage when no further re-sharpening and reuse is possible/advisable; the shank portion can be used for a longer period by fixing new tool bits or re-sharpening and re-fixing the tool bits.

(II) Chucking

Chucking operations are similar to Turning but employs multi-spindle Automatic Bar and Chucking lathes for repetitive and batch production of small components in large numbers, from rolled bar stocks. The tool holder is turret type and can hold multiple tools matching with the number of spindles on the lathe. Tools used are of similar types as used on turret lathes or automatic lathes of other types.

(III) Shaping

This process can be carried out mainly for surface shaping on a shaper machine, or by milling on a planing machine, depending on the size of the job piece. The traditional reciprocating type of shaper machines are now outdated in most of the modern work shops (as they are slow, inaccurate, do not give good finish due to single point tool, cannot do non-planar shaping); this process is now being done by various types of milling machines, depending on the type of surface shaping to be done.

Tools used on traditional shaper is a single point tool; but milling type shaper uses multi-point cutters moving on a spindle, mostly horizontal, but vertical spindles are also available.

(IV) Shearing, Punching And Trimming

These operations are generally done on steel plates, angles, bars, flats etc. by using Power (Crank type) Plate Shearing Machines, Power Cropping, Power Shearing Machines and Power Presses. While shearing machines and cropping machines uses knife edge type cutters of various lengths and sizes, for punching and trimming, operations on thinner plates and sheets, single action or double action press dies **(indexed compound dies)** are used. For making knife edges and punches, high-carbon steel bars are used with required heat treatment for hardening and tempering.

(V) Planing

(i) When long and large components are to be machined for surface planing with reasonably good level of surface finish (upto 12-13 microns is possible) and geometrical horizontal parallelism, the process used is called **Planing** and the machine used is a Planer. Since such planing is generally done on larger components (for small components say less than 1000 mm long and width less than say 500 mm, planing can be done on a milling machine or on a horizontal borer using face milling cutters, or can be done on a Plano-Miller as well). To do planing on a large work piece (large breadth and length), Double Column Planing machines with single tool or double tool positions are employed with reciprocating type table on which the job is mounted and clamped.

(ii) Tools used are generally single point high-speed carbon steel tools.

(iii) On a Planer Machine, a suitable grinding attachment can also be fitted on the cross rail for grind-finishing the planed surface for better surface finish.

(iv) Size of Planer Machine may vary (depending on jobs to be done) from table size 1500 mm × 750 mm to 10000 mm × 3000 mm (manufacturers can manufacture bigger sizes depending on customer's requirement).

(VI) Facing

The process of facing is applied to obtain a smooth or even contoured face of a component. For small components, this can be achieved by using a general purpose vertical milling machine. For large **non-circular** faces/surfaces, requiring high finish, facing can be done on a **Horizontal Boring Machine** using face-milling cutters of appropriate sizes. For large circular components, (like large dia discs of gas turbine rotors), requiring high accuracy and finish, suitable (preferably CNC) **Facing lathe** should normally be used , not only for quality of finish but also for reduced time cycle.

(VII) Milling

(A) Like the turning, **milling operations** are one of the most commonly adopted machining processes in a Work-Shop, either for manufacturing of new product components or for replacement-components for spare parts of machines and equipment.

Milling is a method of removing metal by a rotating cutter while the work piece remain fixed on the work- table which can move horizontally in X or Y axes depending on machining requirements.

(B) A variety of milling machines are now available in a variety of sizes and configurations including the CNC versions.

The general purpose milling operations may be carried out for:

- Surface milling, Face milling,
- Groove and Slot milling, key-way milling;
- Form-milling (T-slots, Fur-Tree grooves/slots) and profile milling.
- Angular milling, tapered milling.
- Milling inside bores (boring operations)
- Precision Milling (as in **Tool Room** operations, high accuracy parts etc.)

(C) A few of the **commonly used Milling Machines** for the above mentioned milling operations are mentioned below:

*(1) Conventional General Purpose Bench Milling machine.

*(2) Conventional General purpose Horizontal Milling machine.

*(3) Conventional General purpose Vertical Milling machine.

*(4) Conventional General purpose Universal Milling machine.

*(5) Precision Milling Machines (for tool-room applications).

NOTES:

(a) Universal Milling Machines are basically horizontal milling machines with a swivelling table; these machines also can have a universal dividing head, thus capable of doing, besides general

purpose milling, the milling of helices of gears, milling cutters, drills and end mills.

(b) All the above machines are Knee-Type with the working table mounted above the knee structure and with horizontal traverse].

(6) Profile Milling Machines.

(7) Swivel type Milling Machines.

(8) Ram type Milling Machines.

(9) Bed type Milling Machines.

** (10) Special Purpose Milling Machines.

** (11) Planer type heavy duty Milling Machines which are generally called the **"Plano-Milling Machines";** these can also do large and small key way milling operations, besides plano-milling/surface milling operations.

(12) Tool Room Milling machines (special purpose Milling machines with high degree of accuracy: accuracy from 1 to 5 microns.

*****NOTE: Heavier models can also do boring operations upto certain diameter, with high level of precision.***

(13) Gear Milling Machine (special purpose milling for generation of gear teeth.)

(14) Thread Milling Machines (Special purpose with high degree of accuracy: 1 to 5 microns.)

>>Most of the above machines are available in CNC versions also as per customer's requirements.

(D) Cutting Tools for Milling Operations: Cutting tools used for milling operations are called Milling Cutters which is a rotary tool provided with one or more cutting tooth which intermittently engage the work-pieces and remove material by relative movement of work-piece and the cutter. There are a large variety of milling cutters mostly, multipoint, used for a variety of purposes and finish required.

Considering the construction, relief of the teeth, method of mounting and the purpose or use, the milling cutters are further sub-classified as under:

(i) Solid Cutter, carbide or cast alloy tipped solid cutters, and inserted blade cutters (inserted blades may be of H.S. steel, solid or tipped carbide or cast alloy material).

(ii) By Relief Incorporation: Providing Relief for the cutting edges, Profile Relieving and Form Relieving.

(iii) By Mounting Methods:-Arbor-type, shank type, or spindle mounted cutters.

(NOTE: Please see Chapter-8 for more about cutting tools).

(E) **Varieties of Milling Tools**: Out of the large varieties, a few commonly used milling cutters are mentioned below (for greater details, reader is advised to refer to **Tool Engineers/Mechanical Engineers' Hand Books):**

Straight tooth, Helical-light duty, Helical–Heavy Duty, Helical mill-arbor, spiral mill with inserted blades, plain gear milling, carbide tipped plain mill, side milling cutter of different types (Solid or with inserted blades), straddle mill, solid face mill, face and end mill, staggered tooth grooving, arbor multithread milling, sprocket cutters, gear-hobs etc., threading mill, shaped profile cutters for special jobs (angular profile, inserted bladed face and radius cutters etc.), inserted blade end-mills, die sinking, engraving, slitting, reaming mills etc.

(F) **Attachments for Milling Machines:** These are special purpose accessories attached to or fitted on Milling Machines of relevant size and type for increasing the Operational versatility, productivity, augmentation of range etc.

A few commonly used are mentioned below:

(1) Swivel Head Vertical Milling head:-- used for converting a horizontal milling machine into a vertical set-up.

(2) Universal Milling attachment.

(3) Rotary Milling Head attachment.

(4) Rack slitting attachment.

(5) Thread Milling attachment.

(6) Slotting attachment.

(7) Parking attachment (for job handling/manipulating during setting up and unloading).

(8) Arbors, Machine-vices, Index bases, chucking attachment, dividing heads, rotary tables, precision measuring equipment (Scales, verniers, measuring rods, micro-meters/dial gauges can be attached suitably for on-the-job in-situ measurements of movements/cuts etc.).

(VIII) Copy Milling

*This is a special purpose milling process in which material removal from the work-piece is done to obtain a specific preset shape and profile by copying the same from a template or a model by tracer controlled sensing technology for reproducing irregular or complex shapes, such as dies, moulds, cams, aerofoil surfaces (like turbine blades etc.). **Tracer controlled milling** is done coordinating and synchronising similarity of movement of either the path of the cutter and the tracing element, or the path of the work-piece and the model or the **master.**

***NOTE: Depending on the size of the component, the copying/duplicating may be accomplished by devices like cams and follower mechanism, by pantograph mechanism, or by error-sensing devices involving little or no contact pressure through the tracer stylus or probe.**

***Actually, now a days, electronic non-contact/proximity sensors are used particularly in CNC versions as mechanical tracer technology has now becoming outdated.**

>>**The Milling Machine controlled by error sensing device** or Servo-mechanisms utilizing feed-back, may operate under one of the following systems:

- Manual Hydraulic;
- Automatic Hydraulic;
- Automatic Hydro-Mechanical;
- Automatic Electro-Mechanical:
  - With mechanical contacts;
  - With Electronic –spark contact;
  - With Automatic Tape Control (utilizing a template or a pattern, or model which has been symbolised on the tape).

(IX) Jig Boring

(i) Jig Boring is a special purpose **Tool Room** machining process for manufacturing of precision tools, Jigs, fixtures, templates, masters, models etc. requiring high degree of accuracy in the range of less than 1 micron to 2-3 microns.

To meet this rather stringent accuracy requirements, Jig Boring machines are placed inside an air conditioned room where the ambient temperature and humidity are controlled within the prescribed limits.

This machining set-up combines boring, milling, copy-milling operations. It has a single spindle sliding head mounted on a vertical column over a work table supported on machine bed. The work table is adjustable X-Y axes in the horizontal plane by means of very precision lead screws **(in CNC versions, with ball-screws to avoid backlash/over-forward; in some latest version CNC machines, specially designed roller-chain arrangements with very precision movements have been incorporated in place of ball screws which sometimes create hindrance due to shape distortion/breakage).**

(ii) By precision movement of the work-table, the machine locates the work-piece under the spindle holding the tool. Precision setting of the table (and therefore, of the work piece) is achieved by means of end-measuring rods/scales, inside micrometer and precision dial gauges/indicator.

These machines are used for accurately finishing various holes, corners and surfaces in definite geometrical relationship to one another.

Tools used on these machines are generally precision drills, rose or fluted reamers, or even single point tools (for boring operations). For boring purposes, boring tool can be mounted on a boring head which can be adjusted for eccentric movement of boring tool to change the diameter of the bored hole.

*(iii) **Profile Projection: This is not a machining process, but a quality checking device** based on optical projection of profiles on a lighted screen, with some degree of magnification, for comparing the profiles done on a work piece with that of a template/ master and establish the degree of variation occurred, enabling decision about possible rectification , or rejection , or acceptance. Depending on the size of the component, the screen diameter size may vary from 500mm to 1000mm. The latest models are hooked to microprocessors or mini-computers for accurate measurement and recording, besides visual inspection.

NOTE: Mentioned here along with Jig Boring because Jig boring process output uses this equipment very frequently and regularly for Accuracy checks on components produced.

(X) Boring

In this process of machining, rough and finish-machining are done on an already existing bore/hole, the diameter of which may widely vary from a few centimetres to a few metres. In fact, the boring operation that we are discussing here are mainly for medium and large diameter bores, say from 300/400 mm to say around10000 mm, and component weighing from a few kgs to say over 100 tonnes, and having job heights/lengths from 300/400 mm to around 5000 to 7500 mm ***(NOTE: Boring operations on smaller components can also be done on a lathe , milling machine and even on a Jig boring m/c).***

Boring operations for medium and large components can be carried out on the following machines :

(1) Vertical Boring and Milling Machines (Also Called Vertical Lathes):

- Vertical Borers primarily are meant to do vertical boring operations.
- Vertical borers can also do turning (outside/inside the bores) and milling of circular grooves, surface planing/milling..
- These can be single-column type with a single tools post/ram, or double-column type with two tool posts/rams, depending on the job to be done.
- In a vertical borer, the tool is fixed on the toolpost; tool post can move up and down as well as in horizontally along the cross rail, and can also

be kept in angular position ; tool post/ram is mounted on a cross-rail beam, and the work-piece fastened to the turn-table which rotates; on the turn-table face, T-slots are incorporated to fix up/hold the work-piece by proper fixtures/clamps.

- The cross-rail heights in Double Column borer can be varied by sliding up and down the same along the vertical face of the columns.
- The table in small size machines can be driven by AC motor or DC motor through a gear reducer train; speed control by suitable electrical controls for the drive motors.
- In medium and large machines, the table is driven by DC Motor through a gear reducer train, controlled through Ward-Leonard (or Ward-Leonard-Lenzes) MG set control system .
- **In modern machines Thyristorized Static drive control-system with DC motor or, with VFD (Variable Frequency Drive) AC motors are used for speed and torque controls.**

  (NOTE:DC motors, although costlier and heavier have distinct advantage in speed and torque control since they have better speed-torque characteristics, and speed can be widely varied from zero to maximum within a few seconds, and the direction of rotation can be reversed quickly with high torque level).
- In CNC versions, CNC controls (Sinumerik, Fanuc, Heidenhein, Fagor etc.) for 3-axis/multi-axis(even up to 7-Axes) operations are fitted for better accuracy and productivity.
- Tools used for Vertical Borers are single point tools for turning/ grooving/facing/radius cutting etc;
- **ATC (Automatic Tool Changer)** can also be incorporated for multiple tools and for saving on repeated tool-setting time.
- Special **Grinding Attachment** can be mounted on the tool post/ram to carry out finish grinding of machined surfaces.

(2) Horizontal Boring Machines

- These machines are generally used for medium and large components. In a Horizontal Borer, the work-piece is fixed on suitable static job holders/fixtures, and the cutting tools mounted on the horizontal spindle rotate to do the cutting operation..
- These are powerful versatile machines which can do, in horizontal axis, boring, face-milling, drilling, taping etc. With a proper head, these can also do vertical face milling, deep drilling etc.

NOTE:

The Horizontal Boring Machines come in two broad classes:

- ***Fixed Spindle and Ram-type Spindle.***

In ram type, horizontal spindle can travel forward and back- ward along the spindle axis at 90^0 to the machine vertical column. Forward travel of spindle can be as much as 1600 mm in large machines (without affecting rigidity and spindle axis deflection beyond permissible limits).

- Horizontal Borer can be a fixed column machine or a travelling column machine; in medium and large sizes, the machine is, in most cases, travelling-column type having a Rotary Table in front on which work-piece can be mounted, Rotary Table can have indexed for horizontal turning/angular movement.
- **Boring Bars** of suitable size and length can be fitted on the spindle to machine large bores with fairly good level of accuracy.
- **CNC versions** are available with Controls (Sinumerik, Fanuc, Heidenhein, Fagor etc.); Thyristorized drives (DC or VFD motors);
- **ATC**s are incorporated where frequent changes of tools (on one setting) are required/desired.
- Tools used are many types of heavy duty milling cutters, single point tools mounted directly on the spindle, or on the boring bars mounted on the spindle, face-mills, arbor type drilling etc.
- **A special type of Horizontal Borer called "Stub-Borer"** is also employed for large and deep bores on large and heavy work pieces requiring heavy cuts and higher rates of material removal besides high degree of accuracy. In this machine, instead of a normal spindle , a rotating tool carrier in the form of a **Conical Stub** with a **snout** at the end (where the tool is mounted (single point tool); the snout can extend forward in horizontal axis up to about 2000 mm. In case of the Stub-Borer, due to very special nature of the boring job, ATC is not used, but CNC version (**) is only preferred for the specific jobs to be done (for IP, LP casings of large steam turbines, casings of gas turbines, large gas compressor body casings, etc.).

*****NOTE: Since accuracy required is very stringent for certain components, the machine may have to be kept in a air conditioned room to avoid machine body/parts expansion beyond limits during summer seasons (ambient temp. above 22^0C).***

(XI) Grinding

Grinding is also a very commonly adopted method of material removal for rough, medium and finish surface machining of work pieces (which can be metal or non-metal).Special grinding machines are also used for profile shaping using profiled grinding tools (discs, cylinder etc).

In Grinding process, a bonded abrasive wheel, mounted on a suitable machine spindle, and rotating at high speed, acts as the cutting medium on the surface of the work-pieces (metal or non-metal), hard enough to resist compression under action of the grinding wheel.

In a workshop (production or repair), grinding operations are generally done by using following machines depending on the type of job and the material of the work-piece :

(1) Portable hand-grinding tools (electrically or pneumatically operated):

Used for the purposes of rough surfacing of castings, dressing of weldments on welded parts, contour-finishing of curved surfaces, surface preparation before painting/varnishing/metal coating, surface cleaning.

These portable grinders are also used for rough grinding of forged or cast-parts, sharpening of tool-bits (drill bits, single point cutters), removal of burrs, surface preparation of small parts before metal coating, sharpening of knives, etc. **(where machining accuracy is not a major consideration).**

(2) Cylindrical Grinding Machines:

- For finish-grinding of already turned circular/cylindrical parts (OD): both work-piece and the grinding wheel rotates in opposite directions at high speeds (work piece at a lower speed) and pressed against each other. For accuracy, care is to be taken that the centering of both the work-piece and the grinding wheel is as perfect as possible to maintain parallelity of the centre-lines. Depending on the material of work-piece and the level of finish desired (in microns ranging from 0.8 to 3 microns in wet grinding with coolant under pressure), the size and category of the grinding wheel (grain size on wheel is very important consideration) is selected. The dynamic balancing of the grinding wheel is a must before the same is used, to maintain the accuracy to avoid wobbling.
- A large variety of grinding wheels are available from reputed manufacturers and selection can be made by consulting selection-charts given in the Tool Engineers Hand Books, or obtained even from the tool/abrasive wheel manufacturers.
- The cylindrical grinding machines are generally horizontal axis spindle machines, self-standing on a bed like the small lathes, having electrical motor driven system with matching speed and feed controls. The coolant is impinged continuously under pressure by incorporating a suitable coolant pumping system. CNC versions of these machines are also available for better quality and higher productivity.

(3) Centre Less Grinding Machines:

By construction, a centre-less grinding machine is similar to that of a cylindrical grinder. In this process of grinding, for grinding cylindrical surfaces, the work piece is not rotated between fixed centres as was in the case of cylindrical grinding process. The work-piece is instead supported and rotated between three fundamental machine components: the grinding wheel, the regulating or feed wheel, and the work-blade. While the grinding wheel does the actual grinding, the work-blade positions the work piece between the grinding wheel and the regulating/feed wheel which controls the speed and feed, controls the sizing of the work and also controls the travel of the work-piece in the through-feed grinding regime. **The work piece in this case does not have a fixed centre of rotation, but is held in position by the regulating/feed wheel and the work-blade**.

(4) Internal Grinding Machines:

(A) These machines can have broadly the following three working principles:

(i) Machine in which work piece is rotated in a fixed position while the grinding wheel spindle rotates and reciprocate along the length or depth of the hole being ground.

(ii) Machine in which the work piece is rotated and reciprocated to obtain the required traverse during grinding process while the grinding wheel spindle rotates in a fixed position.

(iii) Cylindrical grinders in which the grinding wheel spindle rotates as well as have planetary movements, while the work piece does not rotate but reciprocate for traverse.

(B) Internal Grinders generally are of two types:

(i) Chucking Internal Grinder: The work piece is held in a chuck or a fixture mounted on a face-plate, or at the end of the work-head spindle. The other operational features will follow any of the above quoted working principles **(4)-A(i) to (iii)).**

In these machines, the accuracy of the bearings supporting the rotating parts are of utmost importance to achieve the required finish accuracy.

(ii) Centre-less Internal Grinder: This is an automatic and precision version of the centre-less grinder for finish grinding of internal walls of hollow cylindrical jobs requiring very high degree of finish in the range of less than a micron to 2-3 microns(e.g. inside bores of hydraulic cylinders, inner-races of ball-bearing rings etc.). The work-piece is held in position by two pressure rolls and one regulating roll, and thus eliminating the use of chucks for holding the job. Before the work-piece is set-up for internal grinding, the outer diameter of the roll must be finished ground with matching accuracy (less than a micron to 2-3 micron). **It is also to be ensured that the outer diameter of the thin walled cylinder does not get distorted under the pressure of the rolls.**

CNC versions are also available with full automatic controls of feeding, setting, grinding and removal of work -piece.

(5) **Surface Grinding Machines:** Surface grinders are used for finish grinding flat surfaces/parts. Depending on the type of grinding wheel or grinding segment used and the material of the work-piece, very high degree of finish can be obtained.

Surface Grinders are generally of two types:

(A) **Planer type surface grinder:** In which the work-table with work piece on it reciprocates under a horizontal spindle grinding wheel. These planer type-surface grinders are further categorised into the **following three sub-classes:**

(i) Machines using a horizontal spindle grinding wheel with its straight or recessed peripheral faces working on the work piece.

(ii) Machines using a horizontal spindle grinding wheel in the form of a cup, cylinder or grinding stone segments (multi-segments) mounted on the wheel.

(iii) Machines using a vertical spindle (at 90^0 to the work-table) wheel using a cup, a cylinder, or segments of grinding stone-pieces attached to it.

(B) **Rotary Type Surface Grinder: (*)**

In which the work-table revolves. These are further categorised into two sub-classes:

(i) Machine using the peripheral straight face of a horizontal spindle grinding wheel;

(ii) Machine using a cup, a cylinder or segmented grinding stones mounted on a vertical spindle wheel.

NOTE: In all the above categories of surface grinders, the work-table has magnetic chuck system of holding the flat jobs while grinding is carried out; as this system has a distinct advantage of avoiding scratch impressed on the job from a mechanical or hydraulic clamping (which also may interfere with the grinding wheel).

The wheels are as usual driven by electric motors with matching controls for speed, feed and continuous coolant spraying on the working surface, besides lubrication of bearings of moving parts.

(6) **Special Purpose Grinding Machines:** These machines are designed and constructed for special purpose applications like crank-shaft grinding, cam shaft grinding, gas turbine disc grinding, automobile and aircraft engine valve grinding , turbine blade grinding, grinding of propeller-hubs, struts etc.

Two commonly used types are:

(i) **Creep Feed Grinders:** One of the special category of grinding machine is Creep Feed Grinding machine (with CNC controls also available) for grind-machining of parts such as doing serration or generating teeth on gas turbine discs, with very hard material on which milling operations may be costly and time consuming. The grinding wheels of proper quality and peripheral shapes can be used to obtain the desired grind-machining profiles with attendant coolant spraying systems and with very fine and accurate control on rate of feed and speed. **Wheel dressing attachment is also fitted on to cleanup and dress the grinding faces/edges after each grinding cycle.**

(ii) **Optical Profile Grinders:** These special purpose grinders are used in Tool Rooms for extremely accurate grinding of small shapes on form and circular cutting tools, carbide cutter tips, templates, profiling masters for copy milling, Lamination dies etc. The grinding operation on these machines is controlled by direct comparison of an optically enlarged image of the work-piece, with a large-scale master drawing placed behind a viewing screen, at magnification of 10:1 to 100:1.

(7) **Thread Grinding Machines:** This is a special purpose machine for precision grinding of threads, particularly where thread accuracy (Pitch and Lead) required is 0.8 to 2 microns: for machine-tool-lead-screws, thread gauges, micro-meters, verniers etc. These machines are preferred to be operated in air-conditioned rooms like the Jig Boring or Jig-Grinding machines. Both- the work-piece and the grinding wheel rotates at different speeds and have relative movements along the Linear Axis of rotation and the movement is very accurately matched with the lead per revolution of the thread being ground.

Besides precise rotation and axial movements of work-piece and the grinding wheel will have the facility of angular tilting to achieve the desired threading angle. These machines can also incorporate the devices for truing and dressing of grinding wheel which is essential to maintain quality and accuracy.

These machines can also have construction features of cylindrical grinding, centre-less grinding and internal grinding machines and may employ single rib or multiple-rib grinding on the threads. These machines can have external, internal and universal threading features.

Automatic versions with CNC system of controls can give substantially higher productivity for headless screws, studs etc.

(8) Tool and Cutter Grinders:

(a) These are very commonly used grinding machines in any machine shop and Tool Room. Grinding and regrinding of tool bits are a regular requirement in such shops.

Tool and Cutter Grinders may be broadly divided into two types:

(i) **Universal Cutter Grinders.**

(ii) **Cutter Grinders:**

While the **Universal Cutter Grinders** can, besides sharpening of tools, also do other precision grinding such as **cylindrical, internal, face-grinding for very light parts** with special fixtures and job-holders; the **Cutter-Grinder** type is used only for sharpening of cutting tools.

(b) Prior to use of Tool Cutter Grinder, a good knowledge of tool geometry and specific sharpening requirements of various tools is a must and engineers and operators must be conversant with the required details **(Consult Tool Engineers Hand Books).**

(9) Truing and Dressing of Grinding Wheels: For ensuring quality and accuracy of grinding, it is very essential to ensure that grinding wheels have true profiles as required. Truing refer to removal of materials from the cutting face (sticking to the grinding wheel face) and restoring the sharpness of wheel face that has become dull or loaded with particles from the material ground. These are achieved by special devices, either separately installed in the shop, or in some cases, such a device is incorporated as an attachment to the grinding machine. Diamond tools are used for truing and dressing, but care has to be taken that the diamond tools are held at proper angle and at proper pressure against the grinding wheel face to avoid incorrect operation and damages (de-shaping) to both.

(10) JIG-Grinding Machines: This is a special purpose highly accurate grinder used for tool room operations (making jigs, dies etc.).This machine has the facility of correctly locating a hole and then grinding the hole with finish accuracy in the range of less than a micron to 1-2 microns; while using the principles of rectilinear positioning (as in the case of Jig Borers), the grinding spindle movement is planetary.

The grinding wheel (which is cylindrical) can be moved accurately into the hole while grinding straight or inclined holes; this machine can also grind contours of simple or complex nature. **CNC versions are also available for better accuracy and productivity.**

(11) Abrasive-Belt Grinding Machine: The abrasive belt grinding is a convenient way of stock removal from metal surfaces, both by dry or wet methods (depending on the type of material and the finish desired). This can also be deployed for polishing of surface by proper selection of grade of abrasive belt and the tension/pressure against the work piece.

(12) Gear Grinding: Gear Grinding process is used for finish machining of gear-tooth after the teeth have been generated by gear-hobbing or gear-shaping process, or for generation of teeth by using formed wheels for heavy stock removal as well as finishing. **Gear Grinding machines in different version can carry out grinding as well as generation of spur-gears, straight bevel, spiral bevel and hypoid gears.**

(XII) Broaching Operations

In broaching method of material cutting, a combined process of roughing and finish machining is done in one operation by using broach tools of different cutting edges and configurations by multiple-passes with each pass removing a predetermined amount of stock by each successive set of teeth. Each successive set of teeth on the broach tool remains in cutting contact with the work-piece groove-material only for a short time, sustains a minimum of abrasive action, and therefore, absorbs a minimum of frictional heat. The distinct advantages of machining by broaching process is speed, economy, high-productivity and simplicity in machining. Machining of grooves, slots, teeth in different angles and profiles both external, and internal is carried out by broaching process with good level of accuracy which by other means of machining like milling, conventional slotting, form milling etc. shall require multiple setting and operations resulting in higher costs and much longer machining time. If the broach tools are made of proper grades of steel and handled carefully, the life of broach tools can be much longer than other type of cutting tools. However, broach tools are costlier and require a very special type of grinding machine (called **Broach Grinding)** for sharpening and re-sharpening of the same; broach sharpening machines are costly and require the services of trained operators..

In general, any machining job which can be carried out by milling, drilling, shaping, slotting etc., can also be carried out by broaching process.

Broaching Machines are available in different sizes depending on the job to be done. CNC versions are also available. Very big CNC Broaching machines are used for profiled slot cutting on large gas turbine discs. Broaching machines are generally costlier as they require larger capacity drive motors and associated gear-trains and precision controls.

(XIII) Sawing

(a) Sawing is very common practice of machining for cutting of blanks, rolled stocks, forged pieces and structural steel or other ferrous and non-ferrous items. Sawing of metal is done by using:

(i) Power Hacksaws (reciprocating blades)

(ii) Power Band Saws (continuous endless band saw)

(iii) Circular Cold Sawing machines (Circular cutter discs with saw teeth profiled at the periphery, normally varies in diameter from 300 mm to 1000 mm).

(b) **Of the three types mentioned above**, power band-sawing is most productive, but cannot be used for very thick/large sections for which circular saws are more suitable as these can cut any metal or non-metal sections efficiently and economically under the following conditions:

(i) Selected machine must have proper cutting power,

(ii) Correct saw blade selected for the job,

(iii) Proper cutting coolant to be used,

(c) Engineers working in the shop are supposed to know as to how to select the proper machine, correct saw blades, and correct sharpening regimes (can consult **Tool Engineers Hand Book); also the machine operating manual supplied by the manufacturer).**

(XIV) Drilling

(A) *This* is a very commonly adopted machining process in any workshop for making holes in metal and non-metal parts/bodies. Drilling process originates a hole at the desired spot or enlarge an existing hole by revolving end-cutting tools called a **Drill**. Most of these drill bits have twisted grooved body with cutting edges at the end.

(B) Depending on the type of material to be drilled, size of the hole, the length/depth of the hole, appropriate type of drilling machine with proper drilling bit is selected and used. Drilled holes can be through and through, blind, vertical , horizontal, inclined and with or without internal threading. Drilling can also be stepped: first for larger diameter holes and then further for smaller diameter holes. Larger drilling holes are also made by using drilling machines prior to boring operations for still larger diameter bored holes.

(C) **A variety of Drilling Machines are available** and selection of required one is done depending on the application, size of the hole, material of the work-piece and the accuracy level desired. Drilling machines can also use tools other than drills, such as facing mill, end mill, counter- boring mill, chamfering , counter sinking, taping and reaming.

Following are a few commonly used drilling machines:

(1) **Bench Drilling machine:** This is a small diameter (say upto 12 mm dia) generally portable vertical type machine meant for doing small shallow holes of average accuracy. While the spindle is driven by electrical motor through a set of grooved pulleys of different diameter (for varying speed) driven by V-belts; these are also called bench type drill-presses.

(2) **Sensitive Drilling machines:** These are manually fed machines (electrical motor-driven, through reduction gear train or by multiple V-pulleys and V-belts) where operator is allowed to change feed rates as the drill encounters soft or hard spots and inclusions/scrap embedding (in the hole) on the work piece. Some of these are also called **drill-presses** and used as a general purpose machines.

(3) **Vertical Column Drilling Machines:** These machines are similar to the above (2) mentioned category, with vertical spindle, but with a heavier and sturdier construction and with power feeding arrangement for doing heavy duty drilling operations; they are available in various models, including CNC versions with wide range of drilling diameter, speed, and feed rates.

(4) **Radial Drilling Machines:** In this version of drilling machine, a radial arm, which is supported on the main vertical column, swings about the vertical column even up to an angle of 270^0. The drill head slides on the arm and can also swivel to around 45^0 in some model. Depending on the size of the machine, the length of the arm can be between 600 mm to 4500 mm. In heavier models, drilling and taping capacities can be upto 160 mm, with powerful motor drive. However, in heavier models while doing high diameter drilling on thick parts, there is a tendency of the drilling head and arm to tilt upwards due to back reactionary force; therefore, care has to be taken to incorporate counter balancing measures in the design and construction of such machine by taking up the matter appropriately with the manufacture at the time of ordering.

CNC versions are also now available for better accuracy and productivity.

Radial drilling machines have the advantage of carrying out drilling operations at multiple locations on the work piece (like tube plates for heat-exchangers, steam turbine condenser plates etc.) with one setting of job and thereby maintaining good level of **hole pitch-accuracy** (which is vital for condenser and heat exchanger plates).

(5) **Gang Drilling Machines:** There are a few models of medium and heavy duty drilling machines which employ independently operating drilling units (with independently working drilling spindles) served by a common work-table and with a common base bed, permitting simultaneous or sequential drilling operations. The units may be fixed in respective locations, or can be independently movable along the work-table.

(6) **Multi-spindle Drilling Machines:** These machines have multiple drilling spindles mounted on one or more heads which drive a number of drills through universal joints and telescoping splined shafts. These machines, also available in CNC versions, are used for drilling of condenser and heat-exchanger tube plates, not only for high productivity, but also for better hole pitch accuracy. In such machines, there are versions available where number of tools can be more than its number of spindles by incorporating a suitable gear-train in the cluster plate mounted on the drilling head.

The drilling range may be from 12 mm to 150 mm depending on size and capacity of driving power.

(7) **Deep Hole Drilling Machines**: These are special purpose horizontal or vertical spindle heavy duty drilling machines to carry out deep drilling operations in thick plates/work pieces, particularly on end plates of very high pressure vessels/heat exchangers, boiler–drums etc., where deep but straightness of hole is an essential requirement. For this operation, either the drill rotates or the work-piece is made to rotate, depending on the shape/size of the work-piece and the design of the machine. In these machines, very high pressure jets of coolant is impinged on the drilling point for cooling as well as dispersion of swarf as the drill proceeds forward. As per latest technology, specially designed drill-bits called the **'Delta Drills'** have through and through holes inside the drill bits and coolant under pressure is passed through the drills, coming out with pressure at the cutting tips, thereby not only cooling, but facilitate quick removal of swarf; this process increases the productivity manifold besides achieving higher quality of finish. CNC versions are also

available with gun drilling and multiple drilling (including Delta process) arrangements. **The drilling range may be from 12mm to 150mm depending on size and drilling power.**

(XV) Taping (for medium and large diameter holes)

These operations can be carried out by drilling machines, horizontal boring machines, even by lathes with special arrangements of tools and tool holding devices. However, this operation is carried out at low rpm (say from 15 to 25 rpm) and therefore, the machine must incorporate system for low speed operations also. **The range may be from 12 mm to 160mm depending on size and driving capacity**.

(XVI) Thread Cutting and Screw Cutting

(1) Thread Cutting:

(a) For cutting of external threads on cylindrical bars/rods, and both external and internal in pipes; threading on cylinders can be done on lathes/turret lathes.

(b) There are special versions of screw cutting lathes which can perform threading, both external and internal, on cylindrical jobs, pipes, rods.

(c) Where precision of thread pitch and thread geometry is very important, the lead-screw on the lathe must be very accurate and machine should incorporate a threading dial gauge which is used to ensure that lead-screw split nut is engaged with the lead screw at the proper moment so that successive cuts can be taken in the same groove or to cause spacing between grooves to enable proper cutting of multiple threads.

(2) Thread/Form Rolling: Threading by cold or hot rolling process for external threading is done on solid bars, rods etc. The work piece is forced rolled between two threaded hard rolls and threading on the work piece is impressed by cold forging method depending on the size and material of work piece; however, in some specific cases, hot-rolling is also done (work piece heated e.g. car/truck wheel axles). There is no material removal in this method of generating threads, and therefore it is not actually a machining process. The threads thus obtained can have very good levels of accuracy like the machined ones depending on quality of the thread rolls used.

(3) Screw Cutting Machines: These are semi-automatic or automatic machines deployed for large-scale production of screws from bars (they are also called **Bar-Auto machines**). These machines are like automatic chucking lathes with arrangements for screw cutting, sizing and cut off , and removal of finished screws. The sizes (diameter and length, screw pitches etc.) can be varied by incorporating devices for changing tools on the turret, traverse and the turning rates.

A large variety of models are available, including CNC versions, for the specific applications and production requirements.

(XVII) Slotting

Slotting operation is like one of the shaping operations and is carried out either by a horizontally reciprocating shaper or by a hydraulically controlled-electrically driven vertical slotting machine employing single point tools mounted on the reciprocating ram. The job is normally mounted on a table which can traverse horizontally and vertically also.

(XVIII) Honing

In this process, stock is removed from metallic or non-metallic surfaces by rather low speed abrasive grinding mostly in cylindrical parts with a rotating abrasive cylindrical tool head having multiple honing stones arranged in vertical positions along the periphery of the rotating tool cylinder. This is actually an abrasive grinding process for accurate finishing inside a cylindrical bore for a true circular facing as well as straightness (vertically)(like the ones done on petrol and diesel engine cylinders, hydraulic ram cylinders etc.). External Honing can also be done on external cylindrical parts like outer surface of ram of hydraulic cylinders. Honing process of abrasive removal of stock from cylindrical surfaces have the following advantages:

- Rapid; minimum of heat generation and distortion.
- Generation of a straight and true round surface by correcting tapers, non-roundness, wobbliness left behind by earlier carried out machining (boring, turning etc.).
- Obtaining any level of surface finish by choosing proper grit sized honing stones;
- Accurate control on size;
- Honing machines can be fixed vertical or horizontal type as well as manually operated portable type.
- Honing sticks are made of grits (Aluminium oxide, silicon carbides, or diamond) with bonding material (Vitrified clay or Resinoid) and embedding Air voids to allow porosity in the stick which reduces generation of frictional heat.

(XIX) Nibbling and Routing

(1) **Nibbling:** It is a process of cutting metal and non-metal sheets or plates or sections with shallower thickness to take out parts/ components of

odd and irregular shapes and contours. The cutting is effected by a vibrating cutter with vibrating speed of 300 to 400 strokes per minute. A wide variety of these machines are in use depending on the size and thickness of the work piece and the cutting power required.

These machines have a self-standing frame with a U-shaped body, lower part of which supports the work piece (generally plates/ sheets) on a disc like table and the upper part supports vibrating cutter at its end directly coming over the job. These machines are run by electrical motors with eccentric crank mechanism to create motions with eccentric crank mechanism to create vertical vibrating movement in the cutter (which is basically a circular sharp punch) which operates in conjunction with a circular disc. In this process of shape cutting on plates/sheets, the cutter follows the contour of a template or drawn figure on the work-piece. Before the nibbling process starts, there should be a drilled hole or deep punch on the work piece at the starting point of the contour. While round punch cutter is used for contour cutting, flat punch cutters are used for slot cutting. Nibbling can also be carried out on pipe ends by suitable modification in the work-holding device. This process of cutting involves minimum distortion of metals, avoids internal strains and formation of invisible fracture, burrs; these machines can give high productivity with quality.

(2) **Routing:** It is a process by which softer materials like ply board, aluminium sheets, plastic sheets, linoleum sheets(and the like) are cut in various regular and irregular shapes and sizes with good speed and clean edges with minimum of burrs and distortions. Routing machines are available in various sizes and models with speed of cutter (which is similar to a short reamer with spiral cutting edges) ranging from 10,000 rpm to 20,000 rpm. While in the self-standing fixed type machine, the cutter is mounted on vertical spindle above the work-table, in the portable version, specially for wood material routing, the cutter rotates on the spindle of a portable hand-held machine with a guide wheel that follow a path along the edges of a special tool guide which may be straight or in large diameter curved path to achieve the desired routing.

Routing machines generally comes in four typical classes:

(i) **Radial Router:** which is similar to a Radial drilling machine, have a movable arm holding the tool at the end and is manually swung over the work piece to the starting point of the cutting.

(ii) **Portable Hand Router:** (as already described above)

(iii) **Profile Router:** with a vertical cutter spindle directly over the work-table.

(iv) **Shaper Router:** In this machine, cutter is held in a cutter head below machine table and the cutter projects above the work table. The work piece is moved over the work-table following a circular guide fixture while the cutting takes place like that of a profile router.

The cutter used for the router machine is a single flute style with either a right-hand (up cut) or a left-hand(down-cut) movement. Spiral right hand router is used on a Radial Routing machine because it can do plunge cutting as it draws up the chips out. Left hand router cutter forces the chips to fall below the work table as it plunges down.

(XX) Gear Hobbing

Gear Hobbing is a kind of milling process by which forged and machined (turned) blanks are cut by shaper type form cutters called a hob cutter. These hobs are generally cylindrical having length greater than the diameter. With one hob, it is possible to cut inter-changeable gears of a given pitch of any number of teeth within the range of the gear hobbing machine. A hob has cutting teeth on the outer side of the worm shaped cylinder which works against the work-piece disc or cylinder, and cutting teeth on the same by removing stock by following a helical path. As the hob rotates in timed-relation with the work-piece blank, each row of successive teeth on the hob cuts the next portion of the gear teeth spaces on the work-piece; speed and feed will depend on the machine's working technology, type of work-piece and the type of gear teeth to be generated.

There are different working principles of hob-cutting depending on the gear hobbing machine design, the type of gear tooth to be generated and the hob-cutter configuration.

Generally, after hobbing operation, the gears are subjected to gear-grinding operation for accurate finishing of gear teeth faces/angles to achieve best possible meshing.

(XXI) Gear Shaping

This is a method of gear tooth generation by the principle of rolling two gears in mesh and where the work-piece is smaller in size and weight. The work-blank, which represents one of the gear element, and rotates, but does not reciprocate, whereas the gear-like cutter with relieved cutting edges, representing the other gear-element, and rotates and reciprocates against the surface of the blank; and thus goes on generating the required tooth profile on the blank. This process of gear tooth generation is quicker and is resorted to for large scale production of racks, cams, ratchets, latches, pinions and many other such similar shaped (regular/irregular) parts. Gear shapers are also available for a wide range of gear-diameters and pitch diameters.

Gear Shapers of suitable models can generate external and internal gears depending on the type of the work-piece and the tooth profiles.

As already stated above, gear-shaper cutters is basically a gear with teeth relieved to provide cutting edges and clearances to facilitate stock removal in the desired pattern.

(XXII) Gear Shaving

This is a process of finish fine machining of already machined gear-teeth (machined by gear milling, hobbing, shaping etc.) to achieve further accuracy on tooth profiles and pitches. In a gear shaving machine the work-piece is rolled/rotated under pressure contacts against three hardened burnishing gears having tooth profiles of the desired kind. However, this process will give successful results only when the work-piece gear had been pre-machined with fairly good level of quality and near to the desired accuracy.

(XXIII) Lapping

Lapping is a final finishing operation on a machined and ground work-piece for (i) extreme accuracy of dimensions, (ii) correction of residual imperfections in shapes, (iii) refinement of surface finish, and (iv) close fit between matching surfaces.

By Lapping Process, the working life of moving/rotating mating parts can be greatly increased by evening out hills and valleys in the mating surfaces.

Lapping can be carried out both manually or by employing semi-automatic machines by using abrasive tools, abrasive belts, abrasive sheets, abrasive grinding pastes, or by cast iron lapping tools –particularly on semi-automatic machines. For certain hard tools like shaving knife, scraper, dies, plug gauges etc.- leather, hard felt and even wood are also used for lapping **(Relevant Charts for lapping process, lap selection, lap finish etc. for different work-pieces/materials given in Tool Engineers Hand Book may please be referred to for proper selection of process and tools.)**

(XXIV) Super Finishing

Super finishing is an abrasive process of achieving a high degree of surface finish which is resistant to further wear and tear and distortion under working. Super-finishing may be carried out on cylindrical surface by using bonded abrasive stick or a cup-wheel for flat surfaces. This process can be carried out by a special purpose attachment fitted on cylindrical grinding machines or on a lathe. There are also special purpose machines to carry out super finishing process on flat or cylindrical work pieces.

(XXV) Polishing and Buffing

(1) **Polishing:** It is a abrasive process of removal of metal from the surface by using a flexible shaft abrasive wheel or an endless abrasive belt with fine abrasive grains working on the work-piece. There are also

pedestal type polishing machines with wheels on the both ends of a horizontally rotating shaft driven by an electrical motor placed in the middle; in fact, rotor shaft of the motor is extended on both ends on which abrasive wheels are mounted.

(2) **Buffing:** Buffing process is carried out on polished or super-finished surface to polish the surface further to give it a shine and a very smooth finish. This is normally carried out by portable or pedestal type machines with fabric wheels made of out of specially treated fabric sheets. For better buff –finish, certain types of buffing compounds are used depending on the work-piece material, along with different kinds of buffing wheels (made of different materials like fabrics, sisal etc.).

(XXVI) Abrasive Cutting

As distinctively different from the process of grinding, the abrasive cutting process is employed for complete severance of a portion/appendix from the main body of a work-piece, by using a sharp edged grinding wheel which has generally coarse and medium grains. This method can also be used for cutting an open-ended slot. While rough cutting off is generally the requirement (such as sample preparation from forged stock, casting pieces, rolled stocks, pipes etc. for metallurgical/destructive/non-destructive tests), by using proper abrasive wheel and a suitable machine (can be a portable or pedestal type) with correct work-piece holding arrangement, accurate cutting off can also be obtained. Depending on the material to be cut, both dry as well as wet (with coolant) abrasive cutting process can be adopted. In some machine components or tooling parts, accurate slots can be cut with high degree of accuracy by using fine-formed abrasive wheels which otherwise has to be done by time consuming and costly milling process. A variety of self-standing manual and semi-automatic abrasive cut off machines are available.

(XXVII) Electro-Machining

(1) Electro-Machining process is employed for very accurate contour cutting on hard steel materials (high temperature alloy steel, tungsten carbide etc.) as these are difficult to cut by other machining processes like milling, shaping, The process is used on normal machinable materials where high accuracy profile/contour cutting is desired. Besides quicker machining and better quality of cuts (which require minimal or no further machining), this is a convenient and economic machining process for large scale production of components like templates, precision form cutting tools, copy masters, precision profiled parts of fixtures etc.

Electro-Machining process can be divided into four groups:

(i) Spark-Over initiated discharge machining,

(ii) Contact initiated discharge machining,

(iii) Electrolytically assisted machining,

(iv) Ultrasonic machining.

The Spark Erosion Machining (SEM), and Electro-Discharge Machining (EDM), processes are being used in production shops, specially for components/parts in Tool Room applications.

(2) A modified version of Electro-Discharge Machining process called the **'Wire Cut EDM'** is now a days being very commonly used in modern Tool Rooms for higher productivity and accuracy in the manufacture of templates, copy masters, profile pads, single point HSS and carbide profiled tools, checking devices for indexing fixtures etc. CNC versions are also available for better productivity and accuracy. In this setup, a series of micro-sparks are generated between the work-piece and a moving wire as the electrodes. Generally positive polarity is given to the wire (brass) and the negative polarity to the work-piece. Wire (brass) diameter ranges from 0.18 mm to 0.5mm and is selected depending on the finish required and the minimum radius in the corners. The sparking takes place under ionised water which keeps the work piece and the wire cool.

It also flushes out (also by thermal shocks at the point of the spark) the eroded electrode and work-piece material from the work area, keeping it clean and spatter-free.

The wire is reeled and fed to the working point continuously from spools by computer-controlled movement in two or more axes simultaneously.

The profile./contour curve data are programmed in an independent off-line programming station and the data is fed to the computer via direct link, or tape, or cartridge, or by cassettes depending upon the system selected by the manufacturer/preferred by the customer..

(XXVIII) Hydro-Jet Machining

Very fine water jets at a very high pressure and great speed is directed through nozzles(small diameter) at the machining point on the work piece, which may be both metallic and non-metallic, for fine cold cutting in the desired contour on sheets, thinner plates to achieve a finish regimen requiring no further machining process. Since there is no heat generation in this process, there no chance of distortion at the cut edge, no burrs, no spattering; can be used for Tool Room jobs.

(XXIX) Machining Centres

The concept of multiple machining (turning, boring, drilling, facing, milling, grooving, copy milling, knurling/threading, grinding etc.) of components in a single setting on the same machine tool was brought into reality and practice through modified conventional machine tools in 1960s by a few reputed machine tool manufacturers in the USA and Europe mainly with the idea of increasing productivity. But with the inherent limitations of conventional machines, particularly for maintaining the desired accuracy in multiple tool-holding, job clamping, sequential indexing and movement-control , the idea did not catch up in a big way till late 1970s when CNC Machine tools with higher accuracy, reliability and productivity were developed with matching versatile CNC controls (Fanuc and Siemens systems mainly), marketed and put into practice in ever-increasing numbers in the modern manufacturing shops in technologically advanced countries. Now-a-days, a large variety and sizes CNC-machining centres, both in vertical and horizontal configurations, are available in the market, many with universal tool heads, and for multi-axis (even upto 7 axis) machining operations on intricate components. Some models of very large CNC Gantry Type Machining Centres (with 5 to 7 axis operation) are available for machining special types of large components like large steam turbine cylinder casings, gas turbine casings, large Kaplan (hydro) turbine blade-profiles, parts of nuclear reactors (calandria, reactor vessel) etc.

Although cost of these machining centres are higher, but in the long run and for regular batch production, they prove to be economic, besides meeting the technological requirements.

In India, the deployment of CNC Machining Centres caught up in early 1980s rather slowly, but by 1990s onwards, their numbers increased year by year, although still in a selective way in a limited number of industries, specially because of high capital cost and constraints of proper up-keep, trouble shooting/repair and availability of spares (particularly CNC spares). Around, 1995-2000 period, the advantages of CNC machine tools and production systems have been increasingly realized even in small and medium sector engineering industrial units, particularly in view of the increasing opportunities in global markets and cut-throat competition for supply of better quality but cost-effective products and components. Many domestic machine tool manufacturers are also increasing their volume of production of CNC machines to take advantage of the increasing demands in industry sector.

As per the latest trends, many reputed machine tool manufacturers in the USA, Europe and Japan are designing, producing and marketing multipurpose and multi-operation (in one setting) machining centres which can turn out finished components from blanks to ready to use stage in one setting, thereby cutting on production cycle time and costs (e.g. steam and gas turbine blades aerofoil portions).

5.2 METAL FORMING TECHNOLOGY AND METAL WORKING PROCESSES

(NOTE: These processes are generally covered under Metallurgical Industries; most engineering goods manufacturing units prefer to off-load their requirements of such items to these metallurgical enterprises for certain advantages like cost-effective procurement, avoiding investments and deploying additional resources, environmental issues (heat, dust, noise etc.)

In manufacturing of machines, structures, product-components etc., there are certain parts or elements which are to be given the required shape and size by processes other than machining and jointing (by welding, bolting etc.). These processes commonly known as the forming, is carried out by pressing, rolling, hammering and bending. Since we are primarily dealing with metal based industrial activities and since the metals used are predominantly steel, mild steel, alloy steel, grey iron castings, forged or rolled steel, we are confining our discussions to forming process for ferrous metals. Let us take one by one in the following paragraphs:

5.2.1 Metal Forming/Metal Working by Pressing, Rolling, Bending etc.

These processes can be used for

(i) Shaping sheet-metals or steel plates, bars, rods by using hydraulic as well as power (crank) presses and relevant forming dies and tools, fixtures.

(ii) Bending bars, structural steel items, pipes, rods, rolled stocks by using hydraulic presses, hydraulic bending machines, lever and wheel combination, manual bending machines for small diameter and small cross-section rods, parts etc.

(iii) Plate Rolling for circular, semi-circular shapes by using 3-Roll or 4-Roll bending machines (hydraulic system with geared electric motor driven or only with geared electric motor driven working systems).

(iv) Plate Straightening by using two-roll rolling machines (geared electric motor driven).

5.2.2 Metal Forming/Metal working by Forging and Extrusion Process

(1) Cold Forging: Certain small and medium sized parts/components made of non-ferrous metals viz- aluminium, brass, copper etc. can be given the required shape and metallurgical/mechanical properties by subjecting the raw material (bars, rods, blanks, etc.) to cold forging process by using high capacity hydraulic presses, since Hot Forging process in these cases results in changing as well as deterioration of metallurgical (grain-structure etc.) and mechanical properties.

(2) **Hot Forging Process:** This process is widely used for ferrous metal (steel, alloy steel) parts/components from small to big size ones where shaping in the desired configuration with high-strength mechanical properties with matching/prescribed metallurgical grain structure is necessary. Depending on the size, shape and material , the hot forging operations can be done in the following ways:

(i) **Hot Forging by Drop Hammers or Counter- blow Hammers:** Pre-heated blanks can be subjected to hammering by pneumatic drop-hammers or counter- blow hammers fitted with matching forging dies (both for envelope forging or open forging). Work piece is handled and manipulated during forging process, by associated manipulator installed near the hammer, or by using hand-held tongs if the weight of the work-piece is smaller.

(ii) **Hot Forging by Hydraulic Press:** Pre-heated blanks are subjected to pressing to give desired shaped by using high capacity hydraulic (or hydraulic-pneumatic combined) forging presses using forging dies. This process is generally used for big size forgings like large crank-shafts, steam turbine and gas turbine rotor shafts, large generator rotor shafts etc. Work piece is handled and manipulated (during the forging process) by EOT–cranes with a special purpose hook-cum manipulator, or by a floor-based manipulator stationed near the press. Envelope forging can also be done for smaller components to be followed by trimming of excess peripheral metal by using a separate trimming press which can be a crank operated-power press.

(3) **Extrusion Process:** This is basically a cold-forging process, using hydraulic presses to give the desired shape with prescribed metallurgical and mechanical properties for parts/components of non-ferrous metals like aluminium, brass, bronze, copper, (and their alloys); small solid and hollow parts, tubes, pipes, small diameter or small cross-section channels, angles etc. can be manufactured with precision dimensional accuracy needing no further machining or mechanical working on these extruded parts.

(4) **The following aspects may be kept in view in respect of Forming Processes:**

(i) Help of Metallurgical engineering experts may have to be taken for selection of proper processes and machinery for the forming jobs. These presses and hammers shall require properly designed and manufactured press-dies and forging dies, and for this, services of the **Designers of Jigs and Tools** are required to design and issue manufacturing drawings to shop engineers (manufacturing) or for facilitating procurement of such dies from external sources.

(ii) The formed parts/components are mostly used for building up a machine or a equipment, or for structures (which may include defence equipment also).

(iii) Selection of proper size and capacity of equipment (presses, rollers, hammers etc.) require expertise in this field.

(iv) Upkeep of equipment and stocking of spares (may be through the maintenance department) have to be organised (in-house or externally).

(v) Selecting and arranging proper dies, their upkeep and storage for possible future use .(NOTE: Jig and Tool designers can be consulted).

(vi) Proper way of die loading and unloading on presses, hammers and fixing/clamping up before operating the press/hammer (improper handling may result in damages to dies which are costly and intricate in construction).

5.3 WOOD WORKING MACHINES

5.3.1 Usage of Wood in Industry

In Engineering industry, wood/timber is mainly used for the following purposes:

(i) Making packing cases for products/components for facilitating safe shipment.

(ii) Making patterns for foundry work (casting), in case a foundry exists in the enterprise.

(iii) Construction works on dedicated buildings/sheds and fittings of doors, windows etc.

(iv) For making furniture items for in-house use.

However, we shall limit our discussions here to the making of packing cases and as this is the most predominant continuous requirement in the engineering industries.

Since availability of wood/timber is becoming more and more scarce and costly, many industries are changing their packing needs to steel containers which are sturdier and safer in transportation and are re-usable. Moreover, most industries today also find it more cost-effective to procure wooden packing cases from sub-contractors (who make them as per client company's specifications and drawings), thereby avoiding investments on wood workshop and operating paraphernalia for this facility.

5.3.2 Wood Working Machines

These machines are mostly like simple metal cutting machines and comes in the following types for specific jobs:

(i) Wood Saws/Band saws;
(ii) Wood planers; wood shavers;
(iii) Wood edge preparation/grooving machines;
(iv) Routers;
(v) Sanders;
(vi) Wood turning machines;
(vii) Wood milling machines;
(viii) Wood Saw sharpening machines.

These machines are manufactured by a class of machine/equipment manufacturers mostly in the small and medium sectors. Accessories and tools are also available from such enterprises and their selling agents.

The basic principles of design and construction of these wood working machines are similar to simple metal cutting machine tools, but since the cutting power required is far less, and also the tools and tool holders are simpler in design and construction, they are cheaper, both in capital as well as operating costs.

■ ■ ■

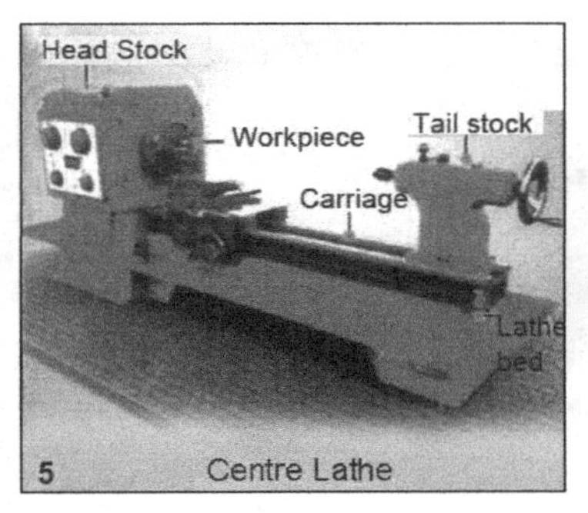

Hydraulic Hacksaw Machine
Cutting Capacity –7" to 24"
3

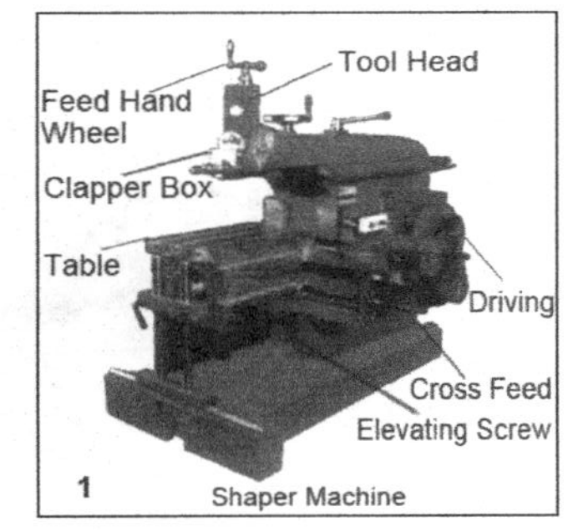

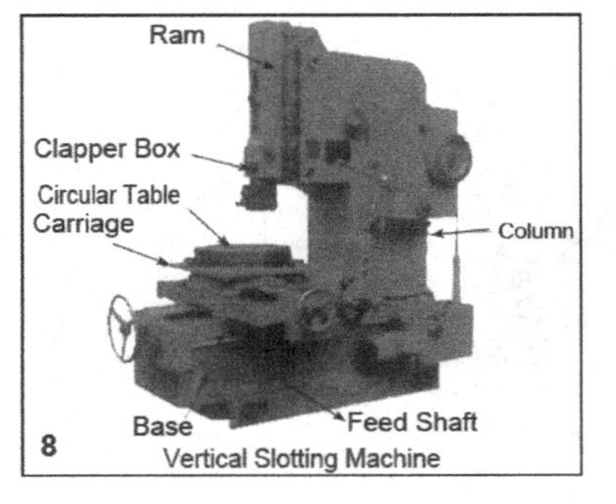

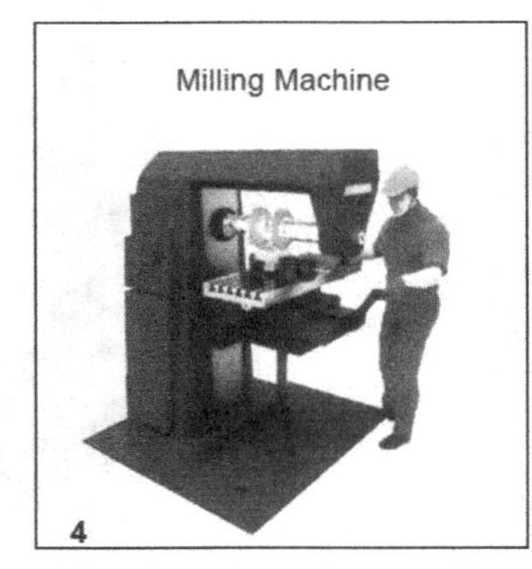

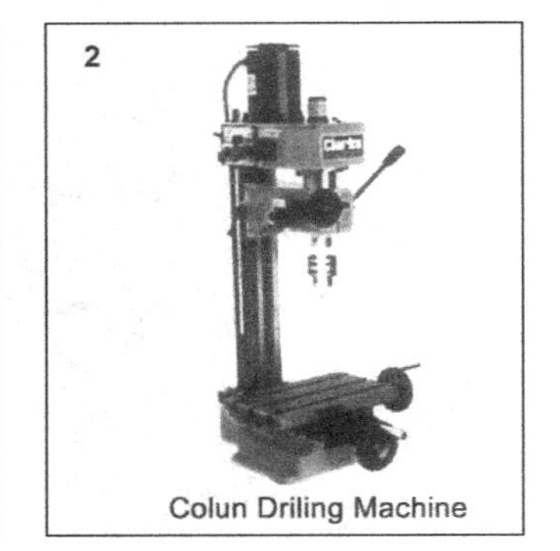

A Representative, Displays Different Kinds of Machine Tools

9 Horizontal Boring Machine

11 Planing Machine

13 CNC Machining Centre

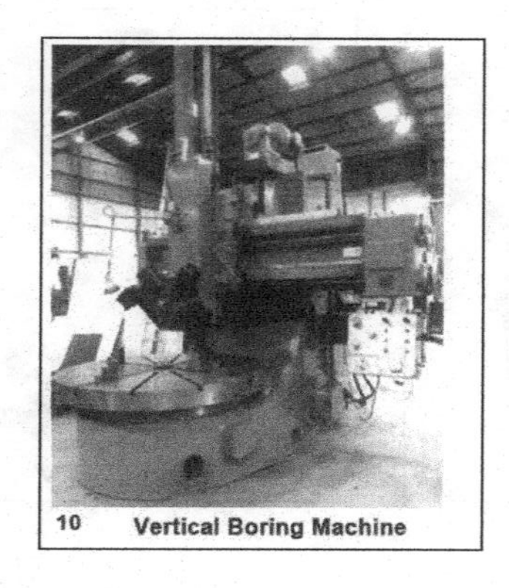
10 Vertical Boring Machine

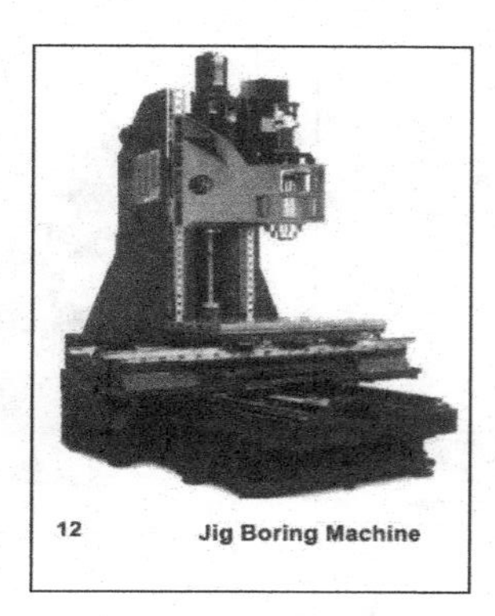
12 Jig Boring Machine

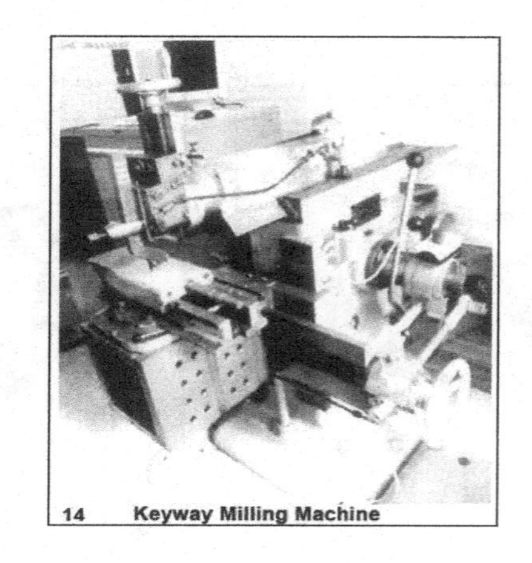
14 Keyway Milling Machine

A Representative, Displays Different Kinds of Machine Tools

6

Techno-Organizational Aspects of Production and Manufacturing Management in Engineering Industries

"If a Production Manager always keeps in mind as to why and for whom he is producing, more than half of his work-stresses may vanish leaving him freer to attend to other nitty-gritty".

—From a lecture delivered by a Management Consultant.

6.0. Production and Manufacturing Aspects

In the Industrial Enterprises, particularly the engineering goods manufacturing units, the role of the **mechanical engineers/technologists and mechanical technicians** are vital for successful and smooth operations, since most of the applicable production technology and manufacturing **activities that take place in the shops are predominantly mechanical process oriented**, from the preparatory and launching stages to the finishing stages. There are however, certain variations in the in--process inspection, testing and measurements for electrical and electronic products and components which may follow a different set of dedicated procedures, process and parameters necessitated by the unique requirements of such electrical and electronic products/components. Nevertheless, irrespective of the industrial unit producing mechanical or electrical or electronic items, the services of the mechanical engineers and technicians are essentially required since they are comparatively better conversant and trained to carry out such shop-floor activities in the specific engineering and technological regimes. Having gone through the topics and issues mentioned in the **Chapters 4 & 5**, it is now the time to enumerate the aspects and issues that the mechanical engineers will be expected to deal with and tackle while working as managers in engineering industrial units:

6.1 Role of the Production Technologists/Shop Process Engineers

They are required to look after and take interest in the following activities with a good level of professional competence and integrity (besides the required skills):

(i) Study and interpret design documents to understand the technological requirements of the products and components to be manufactured. Prepare process documentation to be followed on the shop floor by the supervisors and workers deployed there for product realization through various stages of operations.

(ii) Guide and assist shop engineers, supervisors and artisans in carrying out manufacturing activities as per the prescribed processes and programmes including those for stage-wise inspection and quality control (as per **the Quality Plans and Quality Assurance** documentation issued by the Authorized Agencies).

(iii) Keep themselves conversant/abreast with the technological developments/process improvements that are taking place else where in the specific fields as well as in related activity areas to be able to continuously improve their knowledge and proficiency **not only for better career prospects**, but also enable them to contribute their services in a much better manner to the enterprise where they are working; this may also give them a better job satisfaction which every professional consciously or unconsciously desire. This aspect has also have a bearing on their professional success.

(iv) Keep themselves up-to-date with aspects relating to work–shop practices, machine-shop practices and machining technologies which we have already discussed in details in the **Chapter-5**. Such knowledge inputs are essential for mechanical engineers engaged in production activities.

(v) Keep themselves up-to-date with subjects of Facility Planning for shops and infrastructural requirements and their upkeep so that they can develop capabilities for planning and executing improvements in the manufacturing facilities in case such an opportunity and/ or exigency arises and they are called upon by the authorities/ management to undertake such projects. We shall discuss, more about the subjects of **Facility Planning and Facility Engineering** in **Chapter 9**.

(vi) Keep themselves reasonably conversant with relevant **Standards**, regimes of applicable **Fits, Tolerance and Limits** and also with relevant **Systems and Procedures** in vogue in the enterprise; **a brief on these topics is presented in Charters 7.**

6.2 Production Planning and Control (PPC) Group

(a) This Group functions as the **planning-coordination and monitoring group** for the production department and is normally attached to the Chief Production Manager/Works-Manager. This group prepares and issues detailed production programmes (product-wise, component/assembly-wise, item-wise, etc. with specific calendar schedules e.g. daily, weekly, monthly programmes) to shops engaged in manufacturing activities. This **PPC-Programmes** also tie up action plans and schedules for various inputs to be provided to the shops and coordinate with various support service groups like materials management (purchase, stores, inventory control etc.), commercial and marketing department, quality control department, industrial/productivity services groups, design and engineering department, to ensure timely availability of inputs for the **Production Programmes drawn up and issued.**

The **PPC Group**, besides issuing the detailed production programmes, also draw up and issue **control charts** indicating the **responsibility-matrix** for various inputs, outputs and support services. Such control charts provide for an integrated presentation of activities and calendar schedules linked to the responsibility-matrix for enabling comprehensive management control in the production department.

(b) The **PPC-Group** is also supposed to closely monitor progress on the shop-floors and generate feedback-documents **(MIRS etc.)** for enabling review and control/corrective action by the concerned authorities, primarily from the shop and production management.

(c) In a large production organization having large volumes of physical turnover, there can be major **product-wise PPC-groups** reporting to each product-manager. All these product-wise PPC-groups may get coordinated by a **planning and monitoring cell attached to the CEO of the product-group/Works-Manager as per the system in vogue in the organization..**

(d) The production programmes and control chart formats adopted by PPC groups will be in conformity with the type and complexity of the product and related shop-processes, and therefore may not be in a universalized format; each industrial enterprise has to develop and put into practice formats which are best suited to them; **but all such formats will have the following aspects in common**:

- Names of products, components, assemblies, concerned shops and relevant **work-order number**, product code numbers, part numbers, assembly numbers etc. for traceability and identifications.
- Calendar schedules of execution:-daily, weekly, monthly and quarterly basis **in line with customer delivery commitments.**

- Inputs required at various stages of manufacturing in different shops/activity locations.
- Mention reasons for constraints/bottlenecks if any.
- Indicate responsibility matrix.
- Also mention any critical linkage with other functional groups/ working agency.

(e) **PPC Group** in a way becomes the **'eyes and ears'** of the production manager/works manager/CEO-production (as the case may be). Therefore, personnel posted there will have substantial responsibilities and they cannot afford to be casual or indifferent, nor they should come under undue pressure from the production shops for extending/revising the schedules without adequate justification and due authorization.

6.3 SHOP PLANNING GROUP

Each Area Shop Manager is supposed to have a small planning group to draw up shop-specific action plans based on the PPC-programmes (mentioned above); this cell will also do **intra-shop coordination** and chasing progress and inputs, regularly keep the Area Shop Manager posted with progress and bottlenecks if any arise.

6.4 QUALITY CONTROL AND STAGE-WISE INSPECTION

These are critical functions for the manufacturing engineers and they need to be conscious about the same all the time when working on the shop-floor.

Shop-floor stage-wise inspection and quality control engineers and other inspection staff are supposed to get actively involved in the shop floor manufacturing activities for monitoring quality aspects. **These Quality Control Personnel** are supposed to maintain proper records in prescribed formats for verification and future reference. Engineers and technicians engaged in manufacturing and testing activities must get adequately conversant with the working of the **Quality Circles** and the requirements of **ISO-9001 Quality Management System** irrespective of the fact that the enterprise is already ISO-9001 certified or yet to be certified; future prospects and possibilities are always there, particularly in a market driven competitive business environment, specially in view of increasing globalization of trade and commerce.

- **Adequate knowledge** of application of relevant standards , inspection methodology, and regimes of **Fits, Tolerance and Limits** is also essential for quality control engineers and technicians; a brief on these topics is presented in **Chapters 7.**

6.5 PRODUCT-DESIGN AND ENGINEERING SUPPORT

Product-design engineers extend their support to the manufacturing personnel on the shop-floor as and when necessary; there has to be a good level of mutual co-operation for an unified objectivity based operational culture.

6.6 PRODUCTION/MANUFACTURING MANAGEMENT

Engineers working in production/manufacturing department have great responsibilities towards the organization, since it is their performance, to a large extent, which shall determine the success or failure of the enterprise; they are the real bread-earners of the enterprise. They have to, therefore, organize their works and discharge their responsibilities by taking care of the following **regularly:**

(i) Follow up with marketing/sales department for fetching orders: to load the shops reasonably for optimum utilization of available **resources which include the facilities installed and man-power deployed.**

(ii) Follow up with Product Engineering and Design departments to get design documentation and manufacturing drawings issued well in time in line with the customer orders and customer commitments. These drawings and documents will show configuration, dimensions, material specifications, bill of quantities of materials, other relevant information and instructions, fits and tolerance, **applicable standards etc.**

(iii) Follow up with product-technology groups for technological documentation indicating shop-processes to be followed for fabrication, castings, forgings, machining, forming/working, stages and sequences of shop operations, use of relevant tools, assembly and testing procedures, shop-work process-time norms, use of machine tools, other production equipment and the routing of jobs, (job-card/dockets etc in prescribed formats with relevant details).

(iv) Draw up shop execution plans in line with PPC-programmes and control charts; prepare shop loading programmes for machines/ equipment and for deployment of workers posted their.

(v) Obtain from engineering and/or quality assurance groups applicable **Quality Plans** indicating stage-wise quality checks and norms, also ensuring linkage to relevant standards as prescribed, and ensure further follow up actions.

(vi) Packing: Obtain from Engineering or other concerned departments the details regarding packing procedures and methods, fabrication drawings for packing boxes/containers, raising requisition for standard containers, wagons etc., and ensure follow up actions.

***REMARK: Present trend is to get the packing cases fabricated from sub-contractors to reduce costs and avoid deployment of internal resources for this purpose.**

(vii) Obtain from sales/commercial department procedure for handing over to customer or dispatching to customer by road, rail, air, ship etc. and ensure follow up actions.

(viii) Working out details and taking follow up administrative actions for:

(a) Workers job assignments and deployment as per their skills, trade and as per the service rules.

(b) Identification of training/retraining/skill-up-gradation needs for supervisors, workers/ artisans and taking follow up action with the help of training department (HRDC).(**HRDC= Human Resource Development Centre**).

(ix) Earmarking/defining the roles of shop supervisors, and monitor their performance with respect to the assignments entrusted.

(x) Understanding the role of **Industrial Engineering/Productivity Services Agencies** and maintain liaison/interaction with them as and when required, specially for **optimization of shop-layouts**, workers' performance norm-setting, incentive schemes, productivity projects, processing of suggestions by the employees.

(xi) Getting acquainted with shop procedures for drawing of materials from the main stores, as well as the process of storage in and issue from Sub-stores located in the shops; documents/records must be maintained properly for enabling periodical stock verification / auditing.

(xii) Organizing and managing Tool Cribs and establish procedure for maintaining issue/replacement records.

(xiii) Organizing production planning, scheduling, monitoring and review, procedures/documentation (including scheduling/control charts etc.) thereon. Shop level/departmental level reviews and maintaining records for future reference and follow up actions.

(xiv) Regularly maintaining liaison with engineering, product technology, materials management, commercial/sales/marketing department and also with maintenance and services (for upkeep/repair of machine tools/equipment etc.) department.

(xv) Undertaking coordination and review with shop councils (if any exists) for progress review/monitoring, problem solving etc.

(xvi) Look into implementation of **security and safety procedures** for shop/stores/machines/plants (including fire-fighting arrangements); periodical checking of fire fighting equipment, slings for material lifting, other lifting and dragging tackles/tools, **checking of**

earth-connections and other protective systems of electric supply and distribution networks etc., in conjunction with respective agencies responsible for such services.

(xvii) Dealing with General Administrative matters like establishment, employee welfare, employee grievances, paper work, statutory requirements/reports (as per applicable **Factories Acts/Labour Laws** etc.).

(xviii) Exercising regular vigil for cost and wastage reduction, pilferages, negligence.

(xix) Exercising regular vigil for good house-keeping and for avoiding accidents.*

(xx) Taking special care in handling and storage of hazardous materials on production shops and sub-stores located on the shop floor/attached to shops.

(xxi) Taking care of material handling needs, planning and arranging for the same to avoid holdups and delays.

(xxii) Organizing for availability of special tools and tackles jigs and fixtures, shop accessories **at the appropriate place at the appropriate time**.

* **NOTE:** ***readers may please refer to 5S system described in Chapter 13.***

6.7 OTHER PRODUCTION SUPPORT SERVICES THAT MECHANICAL ENGINEERS SHOULD GET FAMILIARIZED WITH FOR ENRICHING THEIR COMPETENCE AS UP-COMING BUSINESS ADMINISTRATORS

6.7.1 Industrial Engineering (IE), Productivity Services (PS) Functions

Industrial Engineering (IE) department, now also called the Productivity Services (PS) group in some organizations, deals with certain specialized functions to provide support services to the production efforts in an industrial unit. Industrial Engineering group as a separate entity exists mainly in large industrial enterprises with substantial level of physical turnover and manpower. Such large industrial enterprises make use of IE function in many ways to support their operations.

6.7.2

In small size units, **IE** function simply does not exist, nor its role is very much appreciated, whereas in medium size ones, function of IE may be carried out as and when necessity arises, by a small task group attached to the production department, or may be got done by engaging part-time consultant for specific tasks.

6.7.3 The IE department normally deals with traditional IE-functions like

- Work-study, method-study, ergonomics and performance norm-setting,
- Shop layout studies and rationalization efforts;
- Assisting the organization management in formulation and implementation of incentive schemes, working out performance evaluation norms and standards.

>> IE Department may also be entrusted with certain specialized functions such as:

- **Productivity services functions** like helping in identifying and monitoring implementation of productivity- enhancing/savings projects and evaluation of their impacts on organizational performance, both monetary and technological;
- Deal with **Suggestion Schemes** and help in evaluation of the same;
- Provide integrated expert services in formulating **Systems and Procedures** relating to shop-operations; also act **as the catalytic agent** amongst various agencies for an unified systems based approach towards operational performance, thus help in overall functional improvements in the organization.

To know more about IE function and its roles, readers are advised read Industrial Engineering Hand Book Published by McGraw Hill Book Co., U.S.A.

6.7.4 Productivity: A Comprehensive Concept That Helps in Attaining Operational Excellence

"Productivity is the optimized utilization of all available resources, investigation into the best known resources and generating new resources, through creative thinking, research and development and by the use of all possible improvement techniques for the production and distribution of quality goods and services........." taken from a booklet on Productivity Concepts.

(A) Productivity is defined in a number of ways, some of which are presented below:

* Productivity is the ratio between output and input.

* Productivity stands for efficiency in all activities.

* Productivity is elimination of waste in all forms.

* Productivity is the function of providing more and more of everything for more people with less and less consumption of real resources.

(B) European Productivity Council defines productivity as under:

* It is the certainty of doing things better than yesterday.

* It is the constant adaptation of economic and social life to changing conditions.

* It is the continual effort to make improvements.

* It is the faith in human progress.

(C) **National Productivity Council (NPC of India) defines Productivity as under:**

"It is the best utilization of all available resources at our command in order to achieve more and more goods and services with less and less inputs at a desired level of quality to the society and human beings".

6.7.5 Plant Maintenance and Services (PMS) Functions

(A) This is one of the important and sometimes critical function in an Industrial Enterprise for ensuring smooth and trouble free operations of production shop facilities and other related services including infrastructural facilities (mainly Fixed Assets) which comprise of:#

****NOTE: Mechanical Engineers will have a major role in PMS.***

(i) Buildings, Sheds, Structures accommodating production and other related facilities, offices, storage facilities, employee-welfare building facilities (canteen, dispensaries/hospitals, recreation centers, training schools etc.)

(ii) Plant and Machinery such as machine tools, other production machines and equipment, devices etc. that are directly used for production purpose and are installed in the production shops, testing centers, labs etc.

(iii) Services Plants like gas, oil, water supply/ pumping stations, power station, electrical sub-stations, gas plants (oxygen, acetylene, nitrogen, CO_2 etc), telephone exchange, sewage disposal plants, air conditioning plants, etc.

(iv) Services Pipelines for oil water, gas supply systems; under-ground cables and overhead transmission and distribution lines for electrical power supply.

(v) Vehicles, transportation equipment, construction equipment, material handling equipment.

(vi) Roads, bridges, railway siding/yards etc., facilities for river or sea front jetties (in case such facilities exist).

NOTE: Township facilities, if any, is being excluded from the scope of this discussion.

(B) All the above mentioned facilities require regular upkeep and repair-maintenance/reconditioning (as and when required) to keep the same in working order for productive usage. Therefore, there should be a **Plant Maintenance and Services** group in the industrial unit to take care of such activities so that these facilities are available for supporting production activities. As the shop facilities become older with passage of time and being subjected to continuous usage, they start aging and tend to lose their capability and accuracy due to wear and tear, besides going under frequent break-downs/outages thus reducing their availability for productive use. They will need due attention for preventive as well as break-down maintenance and repairs by way of servicing, replacement of worn out/damaged parts (by using spares in stock or by procuring the spares from the manufacturers/dealers/ alternative sources). In severe cases of break-downs/excessive wear and tear, large scale overhauling and/or reconditioning have to be undertaken to restore their operational and productive capabilities and accuracies to the acceptable level.

NOTE: It may not always be possible to restore their original capabilities and accuracies to the fullest extent in spite of major overhauling/ reconditioning because of aging; in that case, a Replacement by a new one may have to be considered.

(C) All the maintenance/ servicing and repair works require substantial organizational efforts and involve considerable resources and expenditures. Therefore, Plant Maintenance and Services group has to be equipped with required facilities, skilled and trained manpower and provided with other resources which should include a well organized repair-shop.

(D) Depending on the size, complexity, type and organizational philosophy of the enterprise, the **Plant Maintenance and Services Group** can either be a separate independent department or can be attached to the CEO of the production management function.

(E) **Maintenance Planning:**

Plant Maintenance and Services department, in order to function effectively and creditably, must keep in focus the following:

(i) Maintaining a well-documented records of all machines, equipment, devices, structures, buildings/sheds, services plants, services lines with all relevant data and information: name, location, layout no., date of procurement, date of commissioning, dates of various repairs and servicing/overhauling/reconditioning carried out, running/utilized hours/days, records of all breakdowns, preventive maintenance done, dates of shifting and re-commissioning, dates of **condemnation(un-serviceable)** etc. as applicable. Machines/Equipment Maintenances Manuals/ Operating Instructions book etc. be properly recorded and kept in a safe central custody. However, for day to day use, a set of copies of such documents may be issued to concerned engineers/ technicians who are supposed to keep these in their safe custody, to be used as and when required.

(ii) Stocking of spares, consumables, repair-tools and tackles.

(iii) Preparing and put into practice preventive maintenance schedules, planned shutdown schedules.

(iv) Maintaining Departmental a Reference Library of machine manuals and other relevant documents.

(v) Keep constant liaison with shops and departments where the facilities are installed and being used.

(vi) Train/re-train the technicians and artisans to upgrade their skills and subject knowledge, **particularly** when high technology new machines/equipment are introduced.

(F) **Maintenance Organization:**

(i) In a large Industrial Enterprise, because of larger numbers of capital machinery and equipment as well as higher level of operational activities (may be in some shops, 3-shifts/round the clock working), the maintenance and servicing activities may have to be organized through functional discipline-based sub-groups **such as mechanical maintenance subgroup, electrical and electronics maintenance sub-groups, civil maintenance sub-groups**. These sub-groups will have to be coordinated and supervised by the **Head of Plant Maintenance and Services department** for a comprehensive approach in maintenance and repair/upkeep to achieve best results.

(ii) **De-Centralized Setups**: In a large enterprise like the one mentioned in (i) above, it may not be practically feasible for the Plant Maintenance and Services departments to deal with all the facilities in use all over the plant in various departments because of resources constraints besides the PMS Setup becoming rather unwieldy to manage effectively.

Therefore, by a considered decision, the division of work and responsibility for upkeep and repair of installed facilities in certain functional areas namely, testing labs, quality control/ inspection, security and fire-fighting services, computer centre, CAD-devices in design department, printing machines, copiers, PCs and other office equipment in various departments, medical equipment in medical department, etc. may have to be dealt with by the user departments themselves, if necessary through AMC (Annual Maintenance Contract) entered into with the concerned suppliers/manufactures or through other competent outside agencies. Since the users are better conversant with the functioning and operational behaviour of the equipment/devices installed/in use in these support services departments, it may be more convenient and comparatively more advantageous if the upkeep and repairs of their own equipment/devices are looked after by the users themselves. Under such an arrangement, Plant Maintenance & Services department will mainly concentrate on the maintenance and repair works on facilities installed in the production shops and for other critical support services areas like equipment in repair shops, vehicular transportation equipment, telephone exchange, vehicular fire-fighting equipment, construction and erection equipment. However, such division of work and responsibilities has to be decided logically keeping in view the organizational set up and deployable resources.

6.7.6 Industrial Safety

In industrial units and also in construction projects, there are areas of activities where hazards/risks exist and accidents may also occur. Therefore, there will have to be conscious and planned safety measures to avoid/minimize the risks of accidents and mishaps. The project management has to, therefore, identify the work-areas where hazardous conditions may exist and which are prone to accidents/mishap and then take preventive/remedial measures. As per the Factories Act and guidelines issued by the Safety Council of India, all such work-premises are supposed to conform to the prescribed norms and procedures. Broadly, the following aspects needs special attention:

(i) **Fire Hazards:** This can occur in almost all the areas due to various reasons including electrical short-circuiting. Fire prevention measures to be taken and fire-fighting implements must be installed in proper places; periodical inspection is imperative.

(ii) **Material Handling:** EOT cranes, suspension cranes, jib cranes, mobile cranes, tower cranes etc. using hooks/eye bolts/U-clamps etc. and slings must be got inspected/tested for safe-loads from time to time, besides taking other safeguards and precaution during their operation as per the prescribed plant safety norms. Damaged/doubtful slings must be replaced with tested and good quality ones.

(iii) **Storage and handling of hazardous materials/chemicals/gases:** Prescribed safeguards and precaution must be taken for these; **'DANGER'** notices must be displayed, besides following prescribed safety norms for storage area.

(iv) **Working at higher levels:** Use to safety-belts must be enforced effectively. Safety belts must also be inspected and tested for safe-load bearing capacity.

(v) **Road Safety:** Prescribed traffic rules and regulations must be followed by all concerned. Vehicles must be got properly checked from time to time particularly for the brake-system, tie-rod linkages and suspension system to avoid/minimize chances of mishaps/accidents. Similarly, all transportation equipment like trailers, wagons etc. must be kept in good and safe working conditions and adequate safety measures taken; periodical inspection of vulnerable parts must be done and if necessary, replaced/repaired.

(vi) **Welding:**

- Welders must wear safety eye-guards, hand gloves.
- Precautionary measures to be taken to avoid chances of fire incidence from welding sparks falling on equipment/materials lying below and around.
- **Foundry Shops**
- Use of protective shoes, gloves, eye-guards, aprons etc. by workers must be enforced.
- Other prescribed safety precautions must be taken.

(vii) **Electrical Installations**

Safety measures and precautions must be taken against:

- Flashovers/sparks/short-circuiting in HV and MV/LV system.
- Proper safety tools and implements must be used by workers while working on live-line systems.
- '**DANGER**' notice must be displayed at the appropriate places.
- Only trained and authorized persons should handle and operate Electrical Installations.

NOTE: Provisions of Electricity Act and Electricity Rules must be kept in view while taking safety measures for electrical installations. Engineers must get familiarized with the provisions of applicable Statutory Regulations.

(viii) **Fuel Storage and Handling:** Prescribed precautions and safe-guards must be taken to avoid leakages, spills, fire hazards, pilferage, sabotage etc.

(ix) **Storage and handling of Explosive materials:** Prescribed safe-guards and precautionary measures must be taken to avoid mishaps, pilferage, sabotage etc. (**NOTE: Provisions of the Explosives Act must be kept in view).**

(x) **Protection against incidence of Lightning Strikes:** All tall buildings, structures, HV and MV Electrical installations, fuel oil storage, other vulnerable installations must have a well laid-out lightning protection system with proper earthing connections.

On over-head HV and MV Electrical Transmission and distribution systems, adequate numbers of Lightning Arresters of proper rating must be installed with adequate earthing.

(xi) All concerned personnel must be properly instructed and trained to use safety implements, fire-fighting equipment etc. also they may be made conversant with measures to be taken during disasters/ mishaps.

Telephone Numbers of Fire-department and those of other safety agencies and emergency services groups must be displayed in all work-stations and vantage points.

(xii) **Organizational Status of Industrial Safety Functions:** Activities of this function predominantly relates to operational areas in the factory premises. By choice of the top management, this function can be attached to the CEO of production department, or with the Plant Maintenance and Service department. Where the Plant Maintenance & Services function gets attached to the CEO of production function, Industrial safety may also get attached to him.

(xiii) Besides working for accident prevention measures, the industrial safety department has to maintain all records of accident, generate and submit **statutory reports** and circulate educative instructions to various departments, display accident prevention posters at vantage points, observe safety week at least once a year.

(xiv) Like quality, safety is also every body's business and **all must consciously follow safety rules and instructions to prevent mishaps/ accidents which can be very costly**

6.7.8 Tool Room Practices

A. Tool Room facilities and the expertise in **Tool Engineering** are essential for any manufacturing work-shop having different types of machine tools and their applications. This is also true for an well-equipped modern mechanical repair-reconditioning work-shop. While large and medium size enterprises prefer to have their own **in-house Tool-Room Facilities** for their work-shops, small scale industries cannot afford to have elaborate tool-room facilities which are costly entailing substantial capital investments, besides having to have the required trained engineers and technicians; also medium size industries may not have all the tool room facilities to work independently. Even in large industrial units, many special items like Jigs, fixtures, templates etc. are being procured from other specialized industries, besides procuring the standard tools from reputed manufacturers.

B. **As regards calibration** of different types of gauges, scales, verniers, micrometers etc., most of the industries depend on outside agencies specialized in this line of activities. For small scale industries, many states have established centralized government owned/sponsored or cooperative Tool Rooms in various industrial estates to cater to the needs of industries around.

C. Mechanical Engineers working in tool rooms and machine shops must be familiarized with the different aspects of Tool Room practices: a brief on the topic presented below for the benefit of the young mechanical engineering professionals:

   (i) To understand the use of various standard tools for metal cutting, metal forming/working, wood working and get to know about their dependable supply sources.

   (ii) Selection criteria for standard tools for various machining/ manufacturing applications (job-process requirements, tool geometry, cutting parameters, rake angles, etc.).

   (iii) Design criteria for non-standard/special purpose tools, jigs, fixtures, dies, templates, copy masters etc.

   (iv) Manufacturing processes for items mentioned in (iii) above.

   (v) Heat-treatment processes for manufactured tooling-items (hardening, case-hardening, salt-bath treatment, tempering etc.)

   (vi) Selection criteria for abrasive wheels, belts, sticks, cylinders etc. for various grinding applications.

   (vii) Re-sharpening procedures for used tools as well as the use of re-sharpening machines/equipment.

   (viii) Storage and conservation procedures for tools, particularly ones with very accurate and delicate cutting edges; conservation methods for long period storage.

   (ix) **Fits and Tolerances** permissible for tools. (consult charts/tables given in Tool Engineers' hand Book).

   (x) **Working Culture**: Tool Room processes involve high degree of accuracy, delicate and skilled working with a conscious sustained efforts and bent of mind for precision quality works; therefore, some care has to be taken for selection of proper personnel to work there; passive and careless workers will make a mess of things and this will have wide adverse repercussions on product quality.

   (xi) Only experienced experts should be entrusted with the task of drawing up purchase specifications and procurement of machine tools for Tool Rooms **(this task is definitely Not everyone's job)**.

Tool Room Engineers, like any other technologists, must constantly keep themselves updated with new developments in Tool Technology, tool-room machines and tool room practices.

(xii) Fitters engaged in Tool Room must have the capability and expertise of **Mill-Right fitters**. Machinists must also be trained properly.

NOTE: For Tool Room's regular production, only experienced and skilled men should be engaged, No trainee nor semi-skilled ones be engaged for regular production/finish production jobs

■ ■ ■

7

Fits, Limits, Tolerances, Engineering Measurements/ Dimensional Metrology

"Not everything that can be counted counts, and not everything that counts can be counted".

–Albert Einstein

7.1 FITS, LIMITS, TOLERANCES

In the manufacturing and work-shop practices, achieving appropriate **Fits and Tolerances** and carrying out proper measurements of dimensions according to the desired/prescribed limits and accuracies, are essential for successful and reliable working of any machinery, equipment, components/parts. Therefore, it is imperative for the dealing engineering professionals to posses adequate working knowledge and appreciation of the relevant definitions, procedures, regimes as well as the use of the related instruments and gadgets to carry out the required checks and measurements so as to be able to determine correctness or otherwise, as well as taking corrective steps as and wherever necessary.

Besides the theoretical and rather limited practical knowledge inputs received from the academic institutions, the young engineers are faced with the task of acquiring actual working proficiencies on an accelerated pace in the industrial work environment to further their competency (and carrier prospects).

It will be useful to regularly consult relevant Hand Books and Standards (Company/National/International) on the subject (ASTME, BIS, BS, Hand Book of Tool Engineers, Hand Book of Industrial Metrology etc.) for finding out prescribed limits of fits and tolerances for various types of components, fixtures, parts etc.

Most industrial workshops make out standard **'Fits and Tolerance Charts'** (based on authenticated sources) and distribute these to the concerned engineers, technicians and artisans for their guidance. Additionally, seeing and doing things on the shop floor as well as regularly interacting with experienced colleagues/coworkers can always be advantageous for increasing knowledge, skill and the confidence level.

(NOTE: More on the subject of application of Standards in Industry is given in Chapter 14).

>> **For the benefit of the young engineers, a brief recapitulation of the relevant topics is presented as under**:

7.1.1 Application of Concepts of Fits, Tolerances and Limits

(i) Machines, equipments, structures etc. comprise of parts/components, sub-units, sub-assemblies which when assembled to form the complete machine, must necessarily be arranged **in a physical and geometric relationship** with each other. The process of jointing faces/surfaces and the matching of dimensions of these surfaces are commonly called **the mating of parts, mating of surfaces and mating of dimensions.** The mutual association of the mating parts within the correct dimensional accuracies result in a **'Fit,'** eventually determined by the accurate operation of the machine. For achieving the **'Fit'** within the prescribed/acceptable limits, the study and determination of accuracies thereof lead to the establishment of **'Tolerance Limits'** and the **'Engineering Measurement'** regimes.

(ii) It is physically impossible to produce two parts which are exactly identical in their dimensional characteristics and even if they are, it is impossible to ascertain it because of the limitations in the Measurement System itself. Hence a certain amount of variability in the required dimensions has to be tolerated. The amount of this variability is known as **'Tolerance'**. It is to be remembered that the tolerance is allowed on parts just because of the limitations in the processes and measurements; otherwise ideally we want the part to have one single identical set of dimensions.

The system of **'limits and fits'** has made the interchangeable manufacture possible. By interchangeable manufacture we mean the production of parts to such a degree of accuracy as is to permit the proper assembly of the functional parts without further machining or fitting, although the individual parts might have been made at different times or in different manufacturing plants/shops on a different set of machines.

The system of Limits and Tolerances have been Standardized the world over and in India too.

(iii) When the machines are to be mass produced, the interchangeability of parts and repeatability of manufacturing process without deviations beyond a certain limits, become essential. However, there can be '**Strictly interchangeable**' parts as well as '**Limited or incomplete interchangeable' ones.** While the **'Strictly interchangeable' parts can** be assembled without selection, adjustments or supplementary operations, the **'Incomplete interchangeable'** parts may require selection or adjustments, but will not require additional fitting or supplementary machining. **'Incomplete interchangeable'** parts, in many cases, enable adoption of the following methods for machining and assembly with higher degree of productivity at reduced costs but with increased tolerance limits:

  (a) Machining to the actual dimensions of the mating parts.
  (b) Machining parts together after their assembly.
  (c) Selective assembly of compatible parts, before final assembly.

(iv) When dealing with **'Fits, Tolerances and Limits'**, one must also have the corresponding knowledge and appreciation of the scope and contents of the related subjects of **'Engineering Measurements'** and **'Dimensional Metrology'**. In the following paragraphs, we shall attempt to briefly discuss the unified role of the above topics so as to enable the young engineers to get acquainted with the related terminologies and relationships and their significance in the design and manufacturing/work-shop practices for machines and equipment.

7.1.2 Engineering Measurements and Dimensional Metrology

"When you can measure what you are speaking about and express it in numbers, you know something about it; and when you cannot measure it, when you cannot express it in numbers, your knowledge is of a meager and unsatisfactory kind. It may be the beginning of knowledge, but you have scarcely in your thought advanced to the stage of a science."

—Lord Kelvin.

(A) **Engineering Measurements:** The role of engineering measurements, with associated science and arts, is an inseparable part of the total process of engineering and industrial activities relating to the manufacturing and work-shop practices. The subject assumes great importance for the mechanical engineers engaged in design, construction, stage-wise inspection, testing and commissioning, as well as repair/reconditioning of mechanical machinery and equipment.

In **Chapter 4,** we have already discussed briefly about various measuring tools and instruments that are generally being used in the work-shops during various stages of activities including the stage-wise inspection for quality control. Since basic theories and procedures for carrying out such measurements, including the use of various instruments/gadgets, are already dealt with in the relevant academic courses and since for the practical applications, appropriate **Hand Books and Standards** are available, we do not propose to go into these details here. However, measurement technologies are constantly undergoing evolutions and improvements with the introduction of electronic/digital, computer/microprocessor and Laser based instruments and devices. There are reputed manufacturers of such instruments and devices, and the interested persons can get the required technical literatures from them.

NOTE: ***Directories of Industrial Products etc, available in Institutional/Organizational/Departmental Libraries, can be consulted.***

(B) **Dimensional Metrology:**

(i) Engineering measurements pertaining to mechanical working systems/machinery get predominantly associated with **Dimensional Metrology**, the science and arts of dimensional measurements.

Dimensional Metrology, therefore, is the scientific process to achieve 'Fits and Tolerances' in the mechanical systems in conformity with the prescribed standards and accuracy regimes.

(ii) **Dimensional Metrology** not only deals with the nature of measurements, like measurements of the magnitudes of Mass, Geometric dimensions, Time, Temperature etc., but also refer to the **Generalized Measurement System** like Metric System, (MKS, CGS) and English System (FPS), followed in various countries/industrial undertakings.

(iii) **Dimensional Metrology** is actually put into practice through devices which may fall broadly into the following categories (**Please also refer to Chapter 4 for more about measuring tools/devices etc.):**

(a) Measuring Tools, Gauges, Gauge Blocks etc.

(b) Indicating Instruments, Indicator Stands, Comparator Stands, Calibration Stands, Standards in physical shapes.

(c) Indicating Instruments for Outside and Inside Dimensions.

(d) Test Equipment and measuring Machines Like Coordinate Measuring Machines (CMMs), Height Measuring Instruments, Universal measuring Machines (UMMs), Gauge Block Calibrator, Millitron, Millipneu, length Measuring Machines, Formaster, Contourgraph, Profile Measuring machine, Micro-tracing System, Universal Measuring Microscope etc.

(e) Thread measuring devices: thread gauge, thread pitch micrometer, optical method on tool room microscope, feeler microscope, knife edges etc.

(f) Optical measurement employing Opto-Mechanical and Opto-Electrical devices: microscopes, Laser-beam position measuring instruments, Optical Profile Projector, episcope boroscope, perthometer, laser alignments telescope etc.

7.1.3 Terminologies Relevant to 'Fits and Tolerances' 'Engineering Measurements' 'Dimensional Metrology

For the benefit of the young engineers, a list of terminologies relevant to the above topics are given below. They are expected to draw upon their institutional (theoretical and practical) /academic input instructions to appreciate full meaning and significance of these:

(i) Nominal size and Basic Size

(ii) Actual Size,

(iii) Mating dimensions,

(iv) Free dimensions,

(v) Enveloped or external mating surface-dimensions,

(vi) Enveloping or internal mating surface dimensions,

(vii) Rounding off of computed dimensions.

(viii) Difference between the maximum and minimum limits is called 'Tolerance',

(ix) The difference between the actual size and the nominal size is called the 'Deviation'

(x) Upper Deviation (UD) and the Lower Deviation (LD),

(xi) Positive Deviation and Negative Deviation

(xii) Base Line corresponding to Normal Size,

(xiii) Tolerance Zone: Upper boundary and lower boundary,

(xiv) Clearance,

(xv) Interference

(xvi) Tolerance of the clearance and Tolerance of the Interference: Tolerances of the 'Fit',

(xvii) Accuracy: dimensional accuracy, accuracy of geometrical forms or micro-geometry of Surface Locational accuracy, accuracy of alignment, parallelity, cylindricity, circular forms.

(xviii) Surface waviness, Surface roughness, surface micro-geometry.

(xix) Errors: Transverse errors, Longitudinal errors, surface errors, locational (relative) errors, systematic errors, constant errors.

(xx) Parallelity of surfaces, Squareness of Surface/planes, Perpendicularity of Surfaces/planes.

(xxi) Economically attainable accuracy, maximum attainable accuracy, Grades of accuracy.

(xxii) Symbols and Notations: Organizational/National/International/Universal) relating to Fits, Tolerances and Accuracies.

(xxiii) Tolerance System (Organizational/National, International Standards) being followed.

(xxiv) Philosophy of and procedure for selection of Accuracy-regimes and Tolerances for different mechanical working systems (by the Designers/Technologists).

(xxv) **Precision**: the repeatability of the measuring process; Precision – versus – Accuracy.

(xxvi) Standards and the Traceability of standards (Sources); Classification of Standards: Organizational/National/International).

(xxvii) Sensitivity and Readability of measuring instruments and devices.

(xxviii) Practicing and adopting Calibration process through in-house accredited facilities, or through:
Nationally/Internationally recognized Calibration
Laboratories-Like national Physical Laboratory (NPL),
New Delhi; Electronic Regional Testing Lab (ERTL),
New Delhi; Instrument Design & Development
Corporation (IDDC), Ambala;
Central Scientific Instrument Organization, Chandigarh;
National Accreditation Board for Testing and Calibration
Laboratory (NABL) etc.

NOTE: Please refer to Annexure 2 for universal measurement units and quantities used in engineering applications.

7.2 FITTING AND ASSEMBLY PROCESSES IN MANUFACTURING SHOPS

These are a set of activities in the work-shop/production shop where components/parts manufactured in different shops or procured from outside are brought into to be fitted into each other and assembled into the desired end-product, prior to dispatch, or prior to works-testing before dispatch (as the case may be). **The engineers assigned to this area must take care of the following for trouble-free and streamlined working:**

(i) Must study and be conversant with the assembly drawings and instructions contained therein.

(ii) Must ensure that workers understand the drawings and know the sequential assembly procedures.

(iii) Ensure availability in the shop, all relevant components, tools, and assembly-materials, in sequential order; regular and close follow-up with feeder shops, other supply-agencies, must be done to ensure this.

(iv) Timely availability of men, cranes, other lifting/material handling equipment.

(v) **In the assembly stage**, there are chances of some degree of mismatching of components which may have to be corrected there itself; therefore services of **skilled mill-right fitters** and assemblers must be made available at the right stage, in right time and in adequate numbers. Sometimes, such activities may have to be taken up at customer's project site premises, and adequate advance planning may have to be done to meet such exigencies.

(vi) **Inspection/Quality control personnel must be asked to carry out stage-wise checks to avoid holdups and delays at the final stage. In fact this should be a regular feature of the work-culture**

7.3 WORKS-TESTING OF PRODUCTS

Depending on the kinds of products and customers' requirements, certain products may have to be works-tested before they are packed and dispatched to the customer. **Engineers posted in this area of activity must take care of the following aspects for ensuring satisfactory performance:**

(i) Must be conversant with the prescribed testing procedures and parameters, including relevant documentation/certification paper work system.

(ii) Must be conversant about the testing facilities and their applications, their capabilities, limitation as well.

(iii) Ensure availability and use of relevant instruments, devices and materials.

(iv) Organize for proper documentation and record keeping.

(v) Ensure that knowledgeable and skilled persons carry out the testing.

(vi) Take care of safety aspects related to the testing process.

(vii) In case customer's representative comes to witness the works-testing, be tactful, polite enough to meet their requirements; in case of a dispute, better refer the matter to a higher authority for resolution/ decision.

(viii) In case testing has been successfully carried out in the presence of the customer's representative, take care to obtain his signatures on the test result sheets/reports generated.

(ix) Take care to involve design and quality control engineers for witnessing the testing and obtain their signatures on the test result sheets.

(x) In case of failures occurring during testing, or unsatisfactory results, the same should be brought to the notice of the superiors immediately for necessary decision before proceeding for further actions.

(xi) In case of a successful testing, organize for systematic dismantling and packing as per documented procedures and instructions.

7.4 SURFACE TREATMENT AND PACKING

(1) Surface Treatments

Prior to packing of various components and parts, some of them are required to be coated or covered with certain paints or varnishes or lacquer or even grease etc. for conservation of their surfaces against rust , corrosion and un-desirable abrasions. Care should be taken to do the surface treatment with the prescribed material only and in the prescribed methods. In case the specified material is not available, authorized agencies such as designers or even the customer should be consulted and their consent obtained before using a substitute.

(2) Packing

Normally, the design of the packing cases required for various components and assemblies, are specified by the product designers depending on the configuration, weight, approximate location of **Center of Gravity**, physical positioning and the structural strength of the component to be packed, **Engineers posted in the area must take care of the following**:

(i) Ensure that correct type and size of packing case is used.

(ii) Correct procedure of packing is followed and recommended internal packaging and shock-proofing materials are used.

(iii) Ensure that the component placed inside the packing case is adequately secured so that it does not roll over inside the packing case during transportation.

(iv) Ensure that a copy of the packing list is placed inside a sealed polythene bag which is then kept inside the packing case. One more copy of the packing list sealed inside a polythene bag is also kept inside a small sheet metal flat box which should then nailed to the outside wall of the packing case **(generally in package No. 1 in case there are more than one package).**

(v) Full address of both the supplier and the customer/consignee should be prominently stenciled on the outside wall of the packing case; destination station should be clearly mentioned. Any other instruction as per customer's requirement must also be imprinted on the packing case. (to the extent practicable).

(vi) Proper marking of loading/slinging points, upright position, protection from rain water etc. should also be indicated on the packing case **(universal symbols should be used).**

(vii) Must ensure timely availability of packing cases and required materials in the packing area to avoid hold-ups and delays.

(viii) If standard steel containers are used, ensure proper size, type, and adequate securing arrangements inside and outside (locking etc.)

7.5 INSPECTION METHODOLOGY FOR MECHANICAL COMPONENTS

Besides the dimensional measurements and metrology, about which we have already discussed above, there are other aspects of inspection methodology which are very important for manufacturing, testing and in-process quality control, carried out through stage-wise checks on tolerances, accuracies, and also on the use of correct category and grades of materials.

For mechanical products and related materials, concerned mechanical engineers are also expected **to get conversant with the scope, characteristics and significance of the following, to be understood in relation to the appropriate standards and the use of appropriate measuring and inspection devices/instruments:**

NOTE: Some degree of metallurgical engineering knowledge is imperative for engineers particularly those in the mechanical, electrical and chemical engineering fields. Even if such subjects are not included in the academic courses, all practicing engineers must acquire some workable knowledge through their own initiatives and efforts.

(i) Metals and their composition (Ferrous and Non-ferrous), classifications.

(ii) Various grades of steel and their compositions (Carbon Steels, Alloy steels, Grey Iron/Cast-Iron, Mild Steel, Creep steel, tools steel, spring steel, boiler quality steel, Inconel steel, etc). **(Please see Chapter 11 on Engineering Materials).**

(iii) **Metal Properties****

Hardness and hardenability, ductility, creep strength, fatigue strength, wear-strength, corrosion resistance, work-hardening, physical/ metallurgical property changes at low and high temperatures etc.**

(iv) **Mechanical Testing:****

Selection of test pieces, tensile test, elongation tests, stress, strain, yield point, yield stress, elasticity, UTS (Ultimate Tensile Strength) etc.

(v) **Hardness tests:****

Brinnel test, Rockwell hardness test, Vickers Hardness test;

(vi) **Fabrication Tests:****

Cupping test, Bend Test, flattering test, drifting test, flanging test etc.

(vii) **Chemical Analysis:****

Determination of contents of carbon, silicon, sulphur, manganese, phosphorous, nickel, chromium, Molybdenum, Vanadium, Aluminum, Titanium, Tungsten, Copper, Tin, Magnesium, Zinc, etc.

****Chemical properties have significant roles in the characteristics of ferrous and non-ferrous metals.**

(viii) **Metallographic Test:****

Micro structures, grain size, grain orientation, inclusions/impurities, cavitations, voids, crack detection etc.

(ix) **Fracture Mechanics and Fracture Analysis:****

The science related to causes of fracture, detection and characteristics of fracture, analysis of the same particularly for mechanical structures, parts which are under mechanical and thermal stresses.

(x) **Heat Treatment Processes and their significance:****

Tempering, hardening, case-hardening, oil quenching, water quenching, induction hardening, Annealing, Soaking, Stress Relieving, Normalizing, Salt-bathing (Cyaniding, Nitriding etc.), carburizing etc.

****NOTE: Consult Handbook of Metallurgical Engineering for details of standard parameters prescribed for various grades of Metals.**

(xi) **Surface Protection against corrosion and protective coating:**

Painting, varnishing, anodizing, nickel/chromium/copper/silver/ cadmium plating, ceramic coating, epoxy coating etc.

(xii) **Welding processes and their quality control regimes:**

(Please refer to para.4.4.1 in **Chapter 4** for details).

(xiii) **Non-Destructive Tests (NDT):***

- Radiographic (X-ray, Gamma Ray);
- Magnetic Particle Tests, Fluoroscopic Tests;
- Ultrasonic Tests;
- Liquid Penetration Tests;

***NOTE: Get familiarized with prescribed testing methods and recommended parameters.**

(xiv) **Optical Checking Methods:***

Employing opto-mechanical and opto-electro-mechnical devices like: Goniometer Ocular, dial-template Ocular, Optical Contactor Level, Optical Profile Projector, Tool Room Microscope, Laser Alignment Telescope etc.

(xv) Surface Finish Checking Methods:*

Employing devices like: Light Section microscope, Interference microscope, Diamond Stylus type surface measuring instrument, Diavite Micro-meter/Surface Roughness Tester, perthometer etc.

(xvi) **Balancing Operations:***

Employing Dynamic balancing and Static balancing devices like Dynamic Balancing Stands, Static Balancing Stands, knife edges, vibrometers, stroboscopes, indicating stands etc.

Computerized/micro-processor controlled devices are now available for accurate measurements and adjustments.

***NOTE: Get familiarized with the prescribed testing methods and recommended parameters.**

■ ■ ■

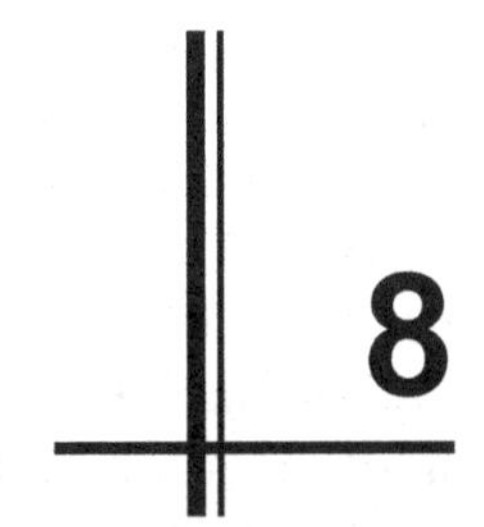

Selection of Machine Tools

"If you keep your eyes and ears open, ready to read a lot on relevant topics, and willing to interact with knowledgeable people, information and knowledge will be at your command."

—Author's view.

REMARK: ***Readers are advised to read books on modern machine tools to know more on the subject. Also reading of technical brochures issued by reputed machine tool manufacturers helps to know about latest trends in machine tool design and technology. Machine Tool Manufacturers' Associations in Europe, Japan and India also furnishes machine tool directories to prospective customer organizations. This also helps in locating supply sources. (Please refer to Annexure-4 listing Industrial Product Directories/Product Finders).***

8.1 SELECTION OF MACHINE TOOLS

Selection of machine tools, either from the existing lots for the ongoing machining tasks, or for procuring new ones, **is rather a complex job** and only experienced engineers/production technologists with adequate in-depth knowledge of manufacturing technology and shop working experience can do proper justice to the task.. While a perfect decision making is a far cry, because of so many variable factors, experienced and knowledgeable engineers should be able to do a reasonably satisfactory job provided they keep themselves updated regularly regarding the latest developments in manufacturing technology, machine tools designs and capabilities, as well as the types/models available in the market. In fact, in the last four decades, there have been a tremendous growth in machine tool markets which have seen a variety of innovative developments in the machine tool-design and configurations for better quality and reliability, higher productivity and accuracy, particularly on the versions with CNC system. Instead of the older concepts of one or two-operations on specific machines, new trendy models incorporating multi-function capabilities are coming into the market, with certain distinctively better features and advantages, as now available in the **CNC Machining-Centres, CNC Machine tools with ATCs, Robotic Operation with Artificial Intelligence (AI) for enabling remote controls, multi-axis operations, application of laser-controls/ measurements, proximity switches for enabling controls without physical contacts.**

Whereas the applications are varied and in wide operational spectrum, we are briefly indicating below, the aspects which young engineers should know and take care in their efforts for enhancing knowledge and proficiency in this field:

8.1.1 For Selecting Machine Tools for Manufacturing Activities in the Existing Shops

(i) Study the product component drawings and the process sheets issued by the concerned technologists/.authorized agencies.

(ii) Understand clearly the material of construction of the component and the physical configuration of the component/work-piece to be machined..

(iii) Understand the sequence of operations prescribed in the process sheets; incase of doubts, get necessary clarifications from the concerned process engineers/production technologists.

(iv) Analyze the machining operations required to be carried out and identify the specific machines on which these operations can be done (like turning, facing, shaping, boring, drilling, copy milling, threading, grinding, rolling, welding, flame-cutting, surface finishing, etc (**please see Chapter 4 for relevant details**), and then plan the loading programme on the selected machines as per the production-planning control charts.

(v) Generate **route cards/job card dockets** (in case such a system is in vogue) with relevant details for guidance of the machine operators.

(vi) Select proper tools/cutting tools, jigs and fixtures, tool holders etc. as per the prescribed procedures/charts (**In case of lack of knowledge in the field, consult tool technologists/Tool Engineers Hand Book in case of doubts/lack of specific knowledge).**

(vii) Before loading the job on the selected machines:

- Be sure of satisfactory operability of the machine and its level of accuracy vis-à-vis the accuracy level prescribed/required for the component.

- Be sure about the tools, tool holders which the machine can take in for facilitating proper machining operations.
- Ensure use/availability of the type of lubricating oils and coolants recommended for the selected machine; alternative grades permissible should also be known.
- Be sure that you have skilled trained operator(s) who can successfully operate the machine.
- Be sure of the trouble-shooting back-up support also and the agency to be approached in case of exigency arise.

8.2 FOR SELECTING OF MACHINE TOOLS FOR FRESH PROCUREMENT (PURCHASE FROM THE MARKET)

(1) For enabling procurement of a machine tool from the market (normally through the purchase department and as per organization's purchase policy), it is essential to draw up a set of comprehensive tender/purchase specifications which must take care of the exact requirements, (immediate and long term.); **keep in view that such machines are purchased once in a while**, depending on the manufacturing/machining technology to be adopted for the current product-mix/work-shop practices, as well as for product-mix/work-shop practices likely to be adopted in the foreseeable future.

(2) Engineers entrusted with such **Facility-Planning (refer Chapter 9 for more on the topic)** and procurement must have adequate knowledge and experience in dealing with such matters, besides keeping themselves abreast with the latest developments in machining/manufacturing technologies, advancements in machine tool design and configurations, and possible supply sources. While selecting and purchasing machine tools from the market (market may offer different varieties for the same types and categories), not only the specifications are important, but also other related aspects like successful installation/commissioning/job-trials, trouble-free operation and maintainability, cost-effective operations etc. are also to be kept in view.

(3) **The following issues are important and must be kept in view:**

(i) Be clear about the various machining processes involved for manufacture/repair-reconditioning of the component/work-piece, from start to finish.

(ii) Be clear about the type and category of the machine tool required to carry out various machining processes for the desired finish/quality, accuracy and productivity.

(iii) Be clear about the types of tools, tool-holders, job-holders/job clamping devices, fixtures etc. that may be required for successful operation and utilization of the selected machines.

(iv) Be sure as to whether you want a '**Single-Operation**', '**Double-Operation**' or a '**Multiple-Operation**' machine (like turning, boring, milling combined, or turning only, or milling only, or drilling only, or grinding only, or you need a combination of many of such operations incorporated in the features of one machining set-up like those available in the **CNC/non-CNC Machining Centres**, horizontal borers, heavy duty universal milling and boring machine and the like.

(v) Draw-up detailed specifications for the basic machine indicating the major parameters like table size, spindle dia, center-distance, swing-over-bed, power of spindle/chuck, weight, and dimensions of the job, drilling dia, taping dia, boring dia, feed rate, speed, rate, accuracy and finish required, traverse required in various axes, bed-length, travel over bed etc;

(vi) Compulsory additional features/accessories required for optimum utilization of the machine being selected..

(vii) Optional features/accessories desired.

(viii) If a **Tooled-up Machine** with dedicated or matching pre-determined toolings is required, clearly specify the types of jobs to be done so that supplier can indicate the types of tools and tool-holders, job-clamping devices required and also recommend other accessories, if any required, for optimal productive use on the machine.

(ix) Types of drives and controls required : AC/DC(thyristor-based), CNC-2-axis/3-axis/5-axis etc. with ball screws, DROs, instrumentation, **cyclic lubrication**, fault-diagnostic devices etc.

(x) Level of finish, accuracy/quality and productivity required.

(xi) **Preferences, if any**, for certain critical items (like make and type of electrical motors, relays, switches, cables, computer/ micro processors, controls, pumps, instruments, CRT, DRO, hydraulic cylinders, hoses, AC-units, oil-coolers etc.) may be clearly indicated, with more than one alternative, for ensuing reliability and reduction in inventory of spares, facilitate inter-changeability and standardization .

(xii) **Vendor** to be asked to quote for commissioning spares separately, and O&M spares for 2-3 years trouble-fee operations separately. (**O&M = Operation and Maintenance**), this will facilitate techno-commercial evaluation of offers easier.

(xiii) **Vendor** to be asked to indicate, recommended grades of Lub oils, coolant etc. (because manufacturer's designers know the best).

(xiv) **Emphasis** be laid on a suitable performance guarantee period: should be at least 12 months from the date of successful trial-run and handing/taking over; for bigger and complex costly machines, it can be any thing between 18 to 36 months (by mutual agreement).

(xv) **Vendor** should be asked to indicate the assured support for after-sales services and trouble-shooting, continued supply of spares. vendor should also be asked to furnish **list of quick wearing parts** with recommended working life; this will **facilitate Maintenance-Planning and Repair/Replacement Scheduling.**

(xvi) Vendor should be asked to furnish list (along with address) of customers where such machine tool is already working.

(xvii) Vendor should be asked to indicate the sources (including full address) of critical bought-out items (fitted on the machine) such as bearings, electrical drives and controls, hydraulic and pneumatic fittings, computers, special devices/instruments etc.

(xviii) Specifications drawn-up should **clearly indicate the acceptance-norms for the successful trial-run/performance tests**.

(xix) Vendor should be asked to indicate the over-all dimensions of the machine, area required for installation and operation, weight of total machine, dimensions of the largest component, weight of the heaviest component; some preliminary details about foundation (if required).

(xx) Vendor should be asked to indicate a definite delivery schedule, schedules for erection and commissioning/job-trials and final handing over; also indicate number of their experts to be deputed for the job and their stay periods.

(xxi) Vendor should be asked to indicate clearly, the role and responsibility of the agent, (if any) in the contractual provisions.

(xxiii) **Purchase/tender specifications should clearly indicate the requirements (no. of sets) of technical documentation including those for machine-foundations, electrical and electronic circuits, hydraulic and pneumatic circuits, ladder-diagram, Installation Manuals, Operations and Maintenance Manual etc.**

(xxiv) Details on training requirements for customer's engineers, technicians and skilled operators should also be asked for.

(xxv) Vendor should be asked to arrange at their cost and responsibility any special tools, devices etc. required by their experts while doing erection and commissioning and job trial works at customer's premises.

(xxvi) **Be sure about the maintainability** of the machine before finalizing the technical evaluation and purchase contract, involve experts from plant maintenances and services department also.

(xxvii) In case of CNC machines, be sure that software being supplied meets the present and the foreseeable future requirements and that these are compatible with the existing set-ups.

(xxviii) Be clear about the level of energy consumption, required environmental/ambient conditions (temperature, humidity etc.) required for trouble-free operation.

(xxix) Be clear about any pollution control measure required to be incorporated in the project/machine, and what will be the additional cost implication (initial and running).

(xxx)* **Take the machine which will serve your purpose with the required level of accuracy and reliability, avoid fancy choices that add to cost without much of an additional advantage.**

Note: Donot over-specify: avoid taking un-necessary extra or sparingly -usable special accessories/ features unless critical for the components to be machined; unnecessary/non-essential purchases lock-up money; you may not get an oportunity to use them effectively; thus these may become dead-inventory over a time and a constant headache for you.

(xxxi) **Carefully weigh the 'Make or Buy' options** for economic operations: need not procure and install machines which will be used once in a while or, will be rather scantily loaded. In case the operation can be got done economically from sub-contract sources, consider this option; **unless critical, absolute self-sufficiency in machining under such** circumstances may not be a prudent policy, as after-all operations/products are to be cost-effective and price-competitive to be saleable in a competitive market.

8.3 USE OF NC/CNC MACHINE TOOLS

NOTE: With the advent of more and more advanced CNC Machine Tools and their many special advantages, NC machines are almost fading out from usage as there is not much demand for such machines any more. However, to fully appreciate and understand the phased development of NC/CNC machines, some knowledge about the back-ground and usage of NC machines is also necessary; hence the same are stated in the following paragraphs.

8.3.1 Historical Background of Development of Numerical Control (NC) Machines

After the **World War-II** ended in 1945, in the years that followed, massive efforts were undertaken in the war ravaged Europe and Japan for re-construction and rebuilding of infra-structures and industries. On the other side, USA , Canada, European countries such as United Kingdom, USSR(now Russian Federation), Germany, France, also embarked on rapid industrial development and modernization efforts for boosting their respective economy and defence preparedness.. These efforts, not only boosted the demands for industrial products and capital goods of various kinds, but also increased the demands for high-productive and reliable quality **machine tools,** specially in the engineering goods manufacturing sectors. The need for developing NC machine tools arose when the complicated profiles of aircraft components could not be produced with consistent quality and accuracy on conventional machines. In case of the conventional machines, the accuracy/quality of the component was totally depended upon the skill of the operator which never remains consistent throughout the operating period. This led to high rate of rejection and hence long delivery cycle for end products. **Therefore, the technologists came to a general conclusion that if consistency of the desired dimension is to be achieved, then the machine should be independent of the skill of the operator**. The other factors which also contributed in fast development of NC Machine were:

- Development of high productivity tools.
- Scarcity of skilled workers.
- Development of computer technology and application software.
- High cost of production.

During the period 1948-52, Mr. John Parsons in collaboration with **MIT, USA**, developed the first NC Milling Machine. However, after some modification, it was announced to the public in 1954.

In India M/s HMT developed the first NC Machine, Turret RAM type milling machine, in 1972.

8.3.2 Application of NC Machine Tools

(1) What is a NC Machine

As the name itself suggests it is the NUMERICALLY CONTROLED MACHINE. Here the word numerical stands for numbers. Therefore, in simple language, the NC Machine means control of machine slides by numbers. The input information is fed through punched paper tape/magnetic tape/MDI (Manual Data Input) in coded language.

(2) Principle of NC Machines

The basic function of a NC machine is to provide automatic positioning of one or more slides simultaneously from one point to another along a prescribed path without the use of cams, drill jigs, templates etc. Thus, NC machine is not a particular machining or forming method, but a better and effective way of controlling machine functions like machine axes movement, spindle speed, feed rate etc., independent of operator's skill.

(3) Classification of NC Machines

(i) Based on the control system features, the NC Machines can be classified as follow:

(a) Point to point NC System;

(b) Straight cut or straight line cut NC System;

(c) Continuous path, or contouring NC System;

(ii) Based on Feed Back Operating System ('Servo System'):

(a) Open Loop System;

(b) Closed Loop System;

(4) Special features of NC Machines

- Simultaneous movement of two or more axes and high metal removal capability of an NC Machine makes it essential to have perfect drive system with rigid structures.
- High positioning accuracy needed in NC Machines makes it essential to have backlash free re-circulating Ball Lead Screw and Nut, and friction free slides: the use of **hydrostatic bearing** system also reduces the coefficient of friction drastically.

(5) Essential Elements of NC Machine

(a) NC console (Electronic Control System).

(b) Machine Tool.

(c) Drive Units.

(d) Feed back system, or servo system.

(e) Electrical Control Cabinet ('Electro-magnetics').

(f) Manual control/operator control pendant.

(6) Mode of Operation

The punched tape (paper tape) is fed to the NC control console where the tape reader reads the instructions which undergo electronic processing. The system senses the electrical commands to the drive units of the Machine Tool and to the electrical control cabinet. Once the machine has commenced its operation, the feed back system sends back reading of the actual position achieved to the NC console where it is compared with the input command and until the difference is brought to zero, the drive unit is actuated by suitable amplifiers from the error signal, to adjust to the preset/desired position.

The manual control pendant allows the operator to perform some of the operations like axes movements, motor start and stop, coolant control, speed change, feed change, editing etc.

8.3.3 CNC (Computer Mumerical Control) Machines

Although NC machines were far advanced as compared to the conventional machine tools, particularly for better accuracy and consistent quality, there are certain operational limitations as can be seen from the comparison given below:

(1) **Major points of difference between NC and CNC Machines:**

NC	CNC
(a) No part programme storage facility and hence every time tape is to be run.	Part programme storage facility is available and therefore tape is run only once.
(b) Data cannot be edited on machine.	Data can be edited on the machine.
(c) No diagnostic facility to indicate any fault or error.	On-line, off-line and resident Diagnostics recourse available to indicate the fault or error.
	With introduction of DNC (Data Numeric Control) System, use of paper tapes and related gadgets are eliminated; and transfer of data becomes easier and more dependable by inter-facing data processing (input - output) between the Central Computer (programming) and the Machine associated Computer.

(2) **To improve upon the operational regimes,** Computer Numerical Control (CNC) Systems were introduced which made the NC machines more versatile for wide-ranging operating capability with many additional advantages. In fact, now a days, CNC machines are more common place in manufacturing work shops than the NC machines. **(Which are getting out of practice and are not much in demand any more).**

8.3.3.1 Application of CNC Machines

CNC Machines are costly machines and therefore their capabilities should be **exploited fully**.

The CNC machines are normally suitable for following types of components:

(a) Components requiring high degree of consistent accuracy.

(b) Components having high rate of rejection on conventional machines.

(c) Components having complex profiles requiring multi axis operations.

(d) Ideal for low to medium batch quantity.

>>Widely used CNC systems on machine tools are:

(i) Sinumerik Systems developed by M/s Siemens AG of Germany:

(ii) Fanuc Systems developed by M/s FANUC, Japan.

(iii) Heidenhein Systems developed by M/s Doctor JOHANNES HEIDENHEIN GMBH, Germany;

(iv) Fagor Systems, developed by M/s FAGOR Automation System of Spain.

The operating ranges depend on application and the type of components; may come in 2-axis to 7-axis configurations, with ball (lead) screws for DC/AC/Universal servo-drives, with DROs, CRTs, Pendant Control Stations with LED based Digital Displays..

8.3.3.2 Advantages of CNC Machines

(a) Rate of rejection is practically nil.

(b) Higher productivity due to simultaneous movement of slides and spindles.

(c) 70 to 80% cutting time **out of total machining time** due to use of programmable speed, feed, tool changing etc.(i.e, substantial reduction in machine setting time before starting actual cutting process),

(d) Higher utilization of machine by way of increased ratio of machining time to non-cutting time.

(e) Lesser and simple holding fixtures required.

(f) Lesser material handling, as most of the operations are carried out at one or two stations.

(g) Flexibility in change of component design.

(h) Reduced inspection time.

(i) Accurate costing and scheduling.

(j) Optimum utilization of capability-(power output) the machine due to higher rigidity of such machines.

(k) Reduction in WIP (Work-in-progress) inventory due to reduction in lead time and faster set-up time.

8.3.3.3 Limitations of CNC Machines

- High cost of the machines.
- Need near perfect pre-planning for their optimum utilization.
- Need very sound trouble-shooting and maintenance team and back-up facilities.
- Not suited for one off component in heavy size as actual saving in time starts after the 1st component, where the programme is established.

- **Difficult to use the machine if the CNC controls are out of order or deficient (cannot be used as a conventional machine even for rough-machining).**

8.4 SELECTION CRITETIA FOR CUTTING TOOL FOR METAL CUTTING MACHINES

(1) **Mechanical Engineers** engaged in production technology and shop operations of various machine tools are required to have some reasonable level of knowledge about the selection criteria and application of various cutting tools. While the tool designers and tool technologies are supposed to have in-depth knowledge and proficiency in their rather specialized field, engineers engaged in actual manufacturing activities should have sufficient working knowledge about the existence/availability, selection criteria and application regimes of the cutting tools for different metal cutting machine tools so that they can exercise a reasonable control on tool consumption, avoid/minimize misuse/abuse and thereby, control wastages and costs.

There is a need to undergo proper training and also read relevant books on the subject (**Mechanical Engineer's Handbook/Tool Engineer's Handbook**).

The understanding of the purpose, tool geometry and cutting regimes for various tools together with shop practices can help in acquisition of knowledge with confidence. The tendency is to depend only on the recommendations of the machine tool manufactures or that given by the standard tool manufacturers for such decisions may be workable, but this approach does not enhance the in-house competency of the concerned engineers. Continued dependence on these external agencies has its implications such as higher operating costs, lack of in-house expertise and flexibility in operation.(particularly in a crisis/critical situation when engineer has to think in terms of alternatives to meet the exigency.)

(2) In **Chapter 5**, we have mentioned about the various kinds of machining process. It can be said that for milling operations, there is the largest variety cutting tools in vogue. Besides the turning operation, milling operation are the most prevalent machining operation in a engineering good manufacturing unit. While the cutting tools used for turning operations are mostly single point tools which are much simpler in design and construction, cutting tools for milling operation require intricate design and manufacturing process for ensuring reliability, accuracy, productivity and longevity as well (since milling cutters are costly).

8.5 ERECTION AND COMMISSIONING OF MACHINE-TOOLS

"Motions and Movements are symbolic of energetic state of life; prolonged static condition tends to freeze energy and kills the urge for productive work."

—Taken from a Machine Tool Manufacturers Brochure

Once the machine is received in the plant, it is very imperative that the same is put to productive use as early as possible so that the machine is utilized for the purpose for which capital investment has been made on the same.

To achieve this, erection and commissioning of the machine be planned well in advance so that as soon as the machine is received, erection works are undertaken, either by in-house efforts (by engaging own resources), or by engaging supplier's experts, or through a third agency specializing in machine erection and commissioning. Whether erection-commissioning is taken up by the purchaser's own personnel or by an outside agency, the purchaser-organization must depute their knowledgeable person(s) to supervise the installation works and get associated at all stages of the project; this is particularly important for the staff who will be operating and maintaining the machine afterwards. In fact, for complex machines, purchaser-organization may depute their personnel to the supplier's works, prior to dispatch, to witness the trial-testing there itself and get training on the operational aspects.

Depending on the level of expertise available, the dispatch clearance (after trial-runs and training is over) can be given by the leader of the same group or the individual trainee duty authorized; or alternatively, customer-organization can engage another inspection-agency duly authorized to carry out inspection and witness the trials (at the manufacturers works); and if found satisfactory, give dispatch clearance (but this option is not a very prudent one).

Engineers engaged in erection-commissioning (and job trials, if necessary for complex machines) are generally expected to take care of the following aspects:

(1) Carry out inspection of the packages of equipment received, if felt necessary, jointly with the representative of the supplier (It will be prudent to do joint inspection to avoid dispute later) to observe and record any damages, short-supply, wrong supply etc. In case of damages, pilferage, at the first instance, notify the insurance company (if covered by a transit insurance policy) and get damage-survey done. At the same time, notify the supplier about damages, shortages or wrong supplies, if any. In case of major damages, obviously, it may not be possible to start erection till replacements are received or damaged parts got repaired. However, incase the damages or shortages are such that erection can be undertaken and completed till the stage of commissioning when replacements/repaired parts are made available.

(2) While for small machine tools, erection commissioning and trial-run **may not be a complex affair**, for the medium, heavy and complex machine tools, a number of steps are involved for carrying out erection and commissioning and job-prove out. These have to be thought of and planned well in advance, and the required resources lined up before starting erection.

(3) Generally, it is prudent to get erection-commissioning, as well as job-proving (if required) done under the supervision and guidance of the manufacture's/supplier's experts so that there is no dispute about satisfactory commissioning/functioning (or malfunctioning) of the machine after or during commissioning/prove-out. This is specially important for medium and heavy/complex machine tools.

>> The following steps should be taken care of:

(i) Readiness of machine foundations-well before machine is received. Check foundation with the machine foundation/layout drawings; carry out any special coating to be applied on the foundation as per recommendation of the manufacturer. Vibration proofing of foundation wherever required/recommended by the manufacturers, should be done as per prescribed method and under expert supervision.

(ii) Carry out mechanical erection/assembly of the machine and check the geometrical alignments of various parts with respect to the bed and with respect to each other (parallelity and perpendicularity, verticality and horizontality etc.). For large and complex machine tools, particularly CNC versions, correct geometrical alignment is very critical for accuracy and rigidity and therefore, special devices like Laser-Alignments apparatus with computerized data analysis are being deployed for checking the alignments (NASA-Tests and calibration system is also executed to ensure accuracy).

(iii) Next is the completion of erection/commissioning works pertaining to electrical, hydraulic, pneumatic and other services lines and sub-systems, and activating them after due checks. Electrical lines/cablings, control panels, switches, relays etc. should be checked, Meggar Tested, and then stage by stage energized/activated and checked for proper functioning. In case of **low insulation level (meg-ohm values)** indicated by the Meggar test, the electrical power lines should be thoroughly checked for any higher-than normal moisture level, earth-leakage, wrong-connections, defective materials/fittings etc.

(iv) Check various drives and controls and their functioning, (electrical and electronics engineers' services may be requisitioned).

(v) Observe machine-noise and vibration levels etc. before taking up or completing final trial-commissioning.

(vi) Let the electrical/electronics/instrumentation engineers check the electronics-systems and instrumentations.

(vii) **After the trial runs are over satisfactorily, then commissioning with job-proving can be taken-up. At this stage be sure of the following:**

- Availability of machine operating manual/instruction documents as well as trained operator(s).
- Availability of work-piece and the process documents/drawings etc.
- Availability proper tools, fixtures, tool-holders etc.
- Availability cranes/job handling devices.
- Check tool setting and job-setting procedures and devices.
- In case of CNC machine, be sure about the correctness of software programmes to be used.
- Involvement of shop engineers/technicians (both production and shop maintenance) in the job-prove out process (to avoid difficulty/ dispute in joint- taking-over of the machine after successful job-trials).
- Keep all operational data generated in proper documentation form with back-up data bank.
- Compare performance (particularly cutting power accuracy and endurance) vis-à-vis those specified for acceptance-norms foreseen in the purchase order specifications.
- Check completeness and relevance of technical documents given by the supplier (including the guarantee certificates).

■ ■ ■

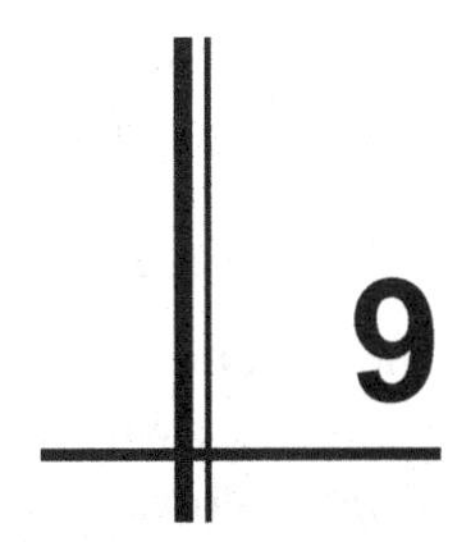

9 Facility Planning for Industrial Enterprises

"Dreaming and Imagination must go together, determined action must follow to integrate both for conversion of the dream into reality to the best extent possible."

—Taken from a lecture on Strategic Planning by the Author

9.1 FACILITY PLANNING FOR INDUSTRIAL ENTERPRISES

All industrial enterprises, small, medium or large, having manufacturing as the core-area activities, must posses certain relevant/dedicated facilities, besides other resources, to support their efforts for turning out saleable products and services to keep them in sustained viable operations.

The Mechanical Engineers have a major role in Facility Planning exercises.

The Facilities can be broadly categorised into two different types by the nature of their utilization and related monetary investments/ expenditures incurred on them.

9.1.1 The First Category comprises of Capital/Fixed Assets type facilities

These are mainly in the form of infrastructural services (roads, power and water supply, gas and oil supply, communication systems etc.), plant and machinery /equipment/devices, land, buildings/sheds, structures, transportation and shipping machinery and equipment, material handling and quality control equipment/devices, testing and laboratory devices, machines and equipment for maintenance and repair services, equipment and devices for design works and printing of drawings etc. (hereafter we may call them as **'Capital Facilities'** since they entail capital investments, and render productive services over a long period(sometimes for decades) with due upkeep and maintenance.

9.1.2 The Second Category of Facilities

These comprise of various shop-accessories of sundry nature, tools, jigs, fixtures etc which are required to support day to day manufacturing activities in production shops, activities in repair workshops and in labs, for quality control etc. These are mostly low cost devices and equipment and these may not be covered under capital expenditure. Expenditure on these may be covered by **operations budget/revenue budget** depending on the company's policy of asset management and financial strategy (within the framework of the Statutory provisions of the industrial and company laws); some of these expenditure may be charged to the Customer Work-Orders depending on the Accounting Policy of the company.

9.1.3 The Capital Facilities

The **Capital Facilities** referred to above in para-9.1.1 entail capital expenditure under Capital Budget and involve substantial funding under the relevant capital investment projects, thus locking up money on a long-term outlay. Therefore, **these capital facilities** are to be judiciously and carefully planned for procurement and installation, keeping a long-term perspective in view.

9.1.4 Facility Planning

Facility Planning activities, especially in medium and large industrial enterprises having products/product-mix involving complex technological processes and multiple functional agencies, assume a regular role to keep the enterprise technologically up-to-date and competitive.

9.1.5

The Tasks involved in carrying out **Facility Planning** activities require the services of experienced and competent engineers and technologists who are adequately conversant with the product configurations (design/engineering, sizing/capacity ranges, application/usage regimes etc.) and the technological processes that are to be followed for manufacturing on the shop floors for which the Mechanical Engineers play a major role. Besides the knowledge on product and associated technological processes, the Facility Planning Engineers/Technologists must also have adequate knowledge about the plant and machinery/equipment/devices already available in the works, as well as that are available in the market, both from domestic and foreign sources, which meet the specific requirements of the particular industrial unit.

9.1.6 The Facility Planning Activity

The Facility Planning activity normally will be undertaken at two stages for the specific industrial unit:

A. **At the first stage,** this will be done at the inception of the industrial project when the Project Feasibility Reports (PFR) in the form of PPR (Preliminary Project Report) followed by the DPR (Detailed Project Report) are being prepared for consideration and eventual approval by the competent authority in the organization;

B. **In the second stage**, which commences after the project is completed and commissioned, certain facility planning activities will go on along with the regular operations of the unit mainly **for the following valid reasons**:

(i) As the unit goes into stream, often certain technological gaps in facility get identified which are needed to be filled to avoid bottlenecks/technological imbalances in operations: which may affect efficiency and productivity, even quality and costs. It is to be appreciated here that with best of intentions and knowledge available, it may not always be possible to draw up a fool proof and comprehensive facility identification plan at the PPR/DPR stages.

(ii) As the unit continue in operations for a number of years, there will arise occasions forcing an upgradation of certain technological processes to meet the changing requirements and preferences of the markets/customers, and this may lead to the necessity of adding new facilities,–may be for better quality and reliability of products, better aesthetics, capacity/size/range rationalisation, adopting new/advanced technological processes for improving competitive edge.

(iii) With continued operations for a number of years, the aging process of facilities starts due to wear and tear resulting in loss of accuracy and depletion in productivity; at that stage, **Replacement** of certain plant and machinery/equipment becomes imperative since these old worn-out items are no more cost-effective in operation because of depleted capability/productivity; rather they become a drag and liability in the smooth operation of the unit. Therefore, a judicious **Replacement-Plan** may have to be drawn up and adopted for implementation to the extent techno-economically feasible and practicable.

(iv) **Technological Obsolescence** is another valid reason which necessitate replacement of machinery and equipment even if they are still healthy and productive, but because of changed technological process (normally more advanced process) they become redundant; therefore may have to be discarded/disposed off before they become a liability and a burden towards the cost of maintenance and upkeep (to avoid their becoming a junk/scrap). The replacement will depend on essentiality of needs and the cost benefit analysis justifying the capital investment decision.

9.1.7 Modernization of Manufacturing Facilities through Reconditioning and Retrofitting

Besides the Replacements and up-gradation as mentioned in paras 9.1.6(ii, iii, iv) above, existing old machines can be reconditioned and even retrofitted with additional advanced features to extend their productive life with the desired level of accuracy and performance, thereby avoiding/reducing capital investments on the new facilities.

9.1.8 The Facility Planning Task Group

A. The facility planning in engineering industries, specially medium and large ones, with complex technological processes, is a task to be performed by a specialist or by a group of specialists with adequate knowledge and expertise. Since collective wisdom is always better in such exercises, the task group may comprise of experienced engineers and technologists from production technology, product design, plant maintenance, quality control and the user departments. The task group may be a semi-permanent cell or a one-time constituted group for a specific project. By the very nature of the scope of the exercises involved, **mechanical engineers/technologists** will have a major (rather leading) role to play in this Task Group.

B. **The Facility Planning Task Group is expected to take care of the following aspects to be effective in their assigned tasks:**

   (i) The Task group should have a compiled directory of all facilities installed in various shops, labs, test-stations, service plants, user departments, duly updated from time to time and with broad specifications, shop-layout/shop-identification numbers, year of manufacture, date of purchase and commissioning/put to use, source of supply etc. These may also be broadly categorised as:

      (a) **General purpose**: production, support services (including plant maintenance and quality control, testing etc),

      (b) **Special purpose**: production, support services (including plant maintenance and quality control, testing, reprographic etc),

(c) **Machine tools**: Semi-automatic, Automatic, CNC, non-CNC, general purpose.

(d) **Services plants/Process plants**: gas, oil, power, water supply, telecom, computer centre, fire fighting, security etc

(e) **Structures** (buildings, sheds, supporting structures, services systems structures etc.).

(f) **Material Handling and Transportation:** equipment, vehicles etc.

NOTE: There can be more elaborate categorization as required for the enterprise's operational needs.

(ii) **The Task Group** will have to have access to historical records of all machines and equipment being maintained by the Plant Maintenance and Services department and also of those maintained by the **user departments, as and when required.**

(iii) **Formal Systems and Procedures** should be in place to enable the user/proposing departments to initiate proposals for procurement/addition of facilities (machines, equipment, devices, buildings/sheds, services lines, structure etc.) with supporting justification so that the **Facilities Planning group can process the same further by way of :**

- Examining and establishing the justification or otherwise of such requests and help in preparing the proposal in the pre-scribed manner. The existing approved systems and procedures (with defined guidelines/parameters) for enabling proper eval-uation of the proposals, be made use of.
- Recommending, if the demand is found to be justified, for **Approval and Budgetary allocation** by the competent authority.
- Help in working out the detailed technical specifications for enabling procurement and installation.

(iv) **The Facility Planning Group** is supposed to regularly collect information on latest technological developments and technical literature along with **budgetary quotations** from the prospective manufacturers/marketing agents for machines, equipment and devices, which are relevant to the areas of operations of the enterprise. This will help the group to build up sufficient level of database/information bank, which shall help in futuristic planning activities, besides enhancing professional knowledge.

(v) **Facility Planning Group** will also play an important role in deciding and processing proposals for re-conditioning and retrofitting cases. In large enterprises, there may be separate multi-disciplinary committees for dealing with and processing of such proposals from different functional areas, but their recommendations are ultimately to be logically processed further by the Facility Planning Task group (**single window processing**) for approval and budgetary sanction by the competent authority and for follow up for actual implementation of such decisions by the concerned executing agencies.

(vi) **Facility Planning Group** shall actively participate in the preparation of the **Annual Capital Budget** since the group is very intimately involved in dealing with **capital investment projects/proposals**. In fact, this group may act as the main coordinating/nodal agency for drawing up and finalization of the capital budget proposals.

9.2 FACILITY ENGINEERING ACTIVITIES IN INDUSTRIAL UNITS

NOTE: Like the Facility Planning tasks as stated above, Mechanical Engineers also have a major role to play in Facility Engineering activities in a running engineering industry.

9.2.1

In a running industrial unit of substantial operational size and having a large number of plant and machinery, and other capital and shop-accessories type of facilities, the predominant emphasis of the enterprise management will be to optimise the utilization of installed facilities and related shop-capacity in order to maximise turnover and profits, subject to of course booking of customers orders to a level to help maintain a reasonable level of operational viability. In such operating conditions, there will be occasions when we need to identify, develop and install special purpose facilities for:

(i) Improvement in /enhancing productivity and optimising utilization.

(ii) Improving quality and reliability.

(iii) Improving ergonomic and safety aspects.

(iv) Reduce energy consumption, fuel consumption and the operating costs.

(v) Undertaking Retrofitting **to improve** operating regimes and technological processes on certain existing machines and equipment.

9.2.2.

While the small and medium size units, because of resources constraints, may not care much to organize in-house efforts for the above activities, larger units which can mobilise resources including lining up competent persons in-house, may like to go in for organizing such activities in-house mainly because:

(i) Cost-effective execution of such tasks (to reduce operating costs).

(ii) To encourage innovative work cultures and development of in-house expertise.

(iii) Self-reliance as well as enhancing technical competency within the organization.

>> **However, it is not to opine** that smaller and medium size units will not like to go in for such improvement efforts, but if need arises, they may hire the services of external agencies to carry out such works provided they are more or less assured about the fruitful results in terms of monetary gains. On the other hand, larger units with progressive outlook generally prefer to encourage such innovative works in-house on consideration of organizational pride also. However, whenever and wherever necessary and cost effective, they may also get such works done through external agencies if reliable and competent ones are available. In any case, for executing such special purpose task, part of the project work like bought–out items (*e.g*: gear box, drive motors, switch-gears, control units, lab units, instruments etc.) as well as certain accessories/ assemblies may be procured from outside agencies as per the specifications drawn up in-house by the concerned engineers/technologists.

9.2.3.

The special purpose requirements mentioned above may not readily be available in the market like the standard design machine tools and equipment with standard specifications. They have to be first conceptualised and then developed to meet specific requirements, and hence require special engineering approach and technological efforts.

9.2.4 Facility Engineering Functional Group

A. Keeping above mentioned objectives and tasks in view, it may be worthwhile for larger units to organize such activities in-house by earmarking fulltime small group/cell of competent and knowledgeable engineers/technologists for the facility engineering functions. Experience in certain large industrial units show that such a group/cell gives best performance if they are attached to the **Plant Maintenance and Services** department mainly because:

(i) The work-culture in the functional area dealing with machine repair-reconditioning and trouble-shooting jobs is routinely oriented to emergent situations forcing the engineers working there to think in terms of innovations and alternative methods to achieve quick and dependable results.

(ii) This department is normally adequately equipped and experienced to undertake such special tasks.

B. Scope of Works for the Group:

The **Facility Engineering Group** (NOTE: In some organization, they call them "Non-Standard Engineering Bureau (NSEB)" is expected to *deal with:*

(i) Drawing up conceptual designs and workout engineering details for the project under consideration; the suggestions may originate from any quarter, but in most probability, suggestion may originate from the user departments, or the production technology department, or even from the machine maintenance department itself.

(ii) Prepare Project Proposal **(Project Report)** for administrative and budgetary approval from the competent authority.

(iii) Prepare working drawing and draw up detailed specifications of materials, components and bought-out items required to execute the project and coordinate procurement.

(iv) Coordinate and supervise/monitor the progress of the portion of the project off-loaded to any external agency.

(v) Ensure eventual completion and commissioning of the project with trial runs and works testing.

(vi) Maintain all records properly.

(vii) Prepare Operating Manual for the equipment/device/setup.

C. To give an idea of the type of projects that can be undertaken by the Facility Engineering/NSEB group, a few examples may be as under:

(i) Retrofit an odd conventional centre lathe with CNC-systems.

(ii) Doing reverse engineering on robotic welding equipment for incorporating such system on a non-robotic welding machine to enhance productivity and quality.

(iii) Providing a roller table with drive for loading and unloading of jobs to a conventional heating furnace to improve on cycle time.

(iv) Providing remote control device on a conventional EOT Crane to avoid whole time deployment of crane operator.

(v) Providing a wall-mounted travelling jib crane using scrap steel material (of dependable quality) and bought-out hoist, to facilitate job loading/unloading on a machine-tool or furnace..

(vi) Providing additional devices (such as extraction fan systems) on a spray-painting booth to remove hazardous fumes quickly and effectively.

(vii) Providing additional shop-automation facilities to improve shop-mobility and reduce waiting time;

– and similar projects to improve shop operations.

D. **Facility Engineering Group** can provide supporting services to **Facility Planning Group** and **Production Technology Group** in executing many of the productivity improvement and technology development projects conceived and undertaken by these groups.

9.3 TECHNOLOGY DEVELOPMENT EFFORTS IN INDUSTRIAL OPERATIONS

"Imagination and Innovation must go hand–in-hand to proceed towards the perceived goal of creativity."

—Taken from a lecture on creativity

9.3.1.

From the topics discussed above under paras-9.1 and 9.2, it is evident that there is a lot of scope for innovative working in an industrial enterprise, especially in large ones operating with high-tech products and processes. In such organizations, there will be enthusiastic engineers, scientists and technicians who will not remain satisfied with routine jobs day in and day out. Given the encouragement and resources, they will like to go beyond routine, think in terms of innovative ideas and develop new/improved processes and devices that can help in:

(i) Improve/enhance productivity of a machine/working system.

(ii) Reduce operating costs (saving in energy, lub oil/ coolant/water consumption, labour etc.).

(iii) Improve ergonomics and reduce fatigue and chances of accidents.

(iv) Improve shop-automation systems and thus help streamline technological flow and reduce/eliminate congestion on the shop floors.

(v) Improve shop-aesthetics and upkeep.

(vi) Inculcate a sense of pride and job-satisfaction in respective work-areas which will have a multiplier impacts on other work-areas as well.

9.3.2.

Experience in some large enterprises shows that such technology development efforts by enthusiastic and talented employees resulted in productivity savings running into crore of rupees in a year (1 to 3% of turnover depending upon the type of product, technological processes, the contributory talents available and the level of encouragements/supports provided by the management). Obviously, such innovative work-culture will need some incentives and reward schemes to be in place to provide continued impetus and encouragement. It is expected that a progressive management will encourage such efforts and provide necessary support and incentives.

9.3.3.

For a comprehensive approach towards an all round improvement in the organizational efforts particularly those organizations who are aspiring for (or are already in possession of) ISO-9001-QMS, TQM and EMS (ISO-14001) Certification, all the three functions as described above (Paras 9.3.1,9.3.2) will have major complimentary and supplementary roles to play besides other organizational efforts and supports from all other functional areas.

NOTE: When we talk about innovative work culture, we are not looking here for earthshaking talents and revolutionary concepts, but are talking about contributions coming from various small conceptual ideas and efforts that may results in tangible and intangible gains to the enterprise besides providing job enrichments and job-satisfaction to those working for the same.

> *"Not all of us have to possess earthshaking talent, just common sense and love will do."*
>
> —Myrtle Auvil

■ ■ ■

10

Mechanical Supporting Equipment and Services Systems in Engineering Industries

"A Large System can operate successfully only when the supporting sub-systems work efficiently in tandem and in sync; every link in the chain is important and must be dependable."

—Author's View

10.1 MECHANICAL SUPPORTING EQUIPMENT AND SERVICES SYSTEM FOR ENGINEERING INDUSTRIES

Besides the machinery and equipment required for manufacturing activities as discussed in the **Chapters - 4, 5 and 8,** there are a number of other mechanical supporting equipment/devices as well as services systems that are also important and in some respect essential for successful and smooth running of industrial units.

In the following paragraphs, we are discussing briefly a few commonly deployed mechanical supporting equipment and services systems that Mechanical Engineers may be routinely called upon to deal with:

(NOTE: Situation demanding, help from electrical, electronics and even civil engineering departments may have to be taken for specific requirements).

10.1.1 Pumps

(1) A variety of pumps are used for pumping liquids, gases/gaseous substances as well as certain semi-liquids such as slurries from Ash and Slag disposal system in Thermal Power Station/boiler-house, sewage from Municipal residential districts.

(2) Pumps are designed and constructed by taking into consideration the following basic parameters/factors:
  (i) Material to be pumped and the site conditions;
  (ii) Discharge rate, operating pressure and temperature;
  (iii) Suction head, delivery/discharge head required;
  (iv) Delivery distance;
  (v) Loading conditions (volume, pressure, temperature, viscosity, physical conditions of the liquid/fluid/slurries to be pumped etc.)

(3) Bodies of pumps are generally made of gray iron castings or steel castings, or alloy-steel castings etc. For high pressure hydraulic oil pumps in small sizes, bodies can be made out of rolled or forged stocks (steel, alloy steel etc.) as well. The impellers/rotors are generally of alloy steel or stainless steel, or even of non-ferrous metals like brass, bronze, duralumin etc., depending on the fluid to be pumped and the loading conditions. When the pumps are meant to pump corrosive material (like acids, chemicals) or abrasive material (like Ash and Slag slurry), the inner surface of the pump body casing may be lined with anti-corrosive/anti-abrasion material like, rubber, polythene, Nylon, Teflon, FRP, or even by ceramic compounds/tiles (***NOTE:- for special application in chemical/drug/food-processing industries, pumps made of ceramic material, Nylon, Teflon are also being used***). In such pumps, the impellers are generally made of stainless steel or even nylon or Teflon (for smaller pumps) etc.

(4) **Pumps can be broadly divided into following categories:**
  (i) Reciprocating type (piston) pumps;
  (ii) Centrifugal pumps (radial flow);
  (iii) High pressure multistage pumps (axial flow/radial-axial flow);
  (iv) Rotary pumps/Gear pumps (generally radial-axial mixed flow);
  (v) Jet pumps for shallow wells;
  (vi) Deep well submersible pumps;
  (vii) Vertical-axial (turbine) pump
  (viii) Sump-pump (can be centrifugal, rotary or turbine type);
  (viii) Hand pumps (plunger type and rotary type).

Selection of any of the above shall depend on application and the criteria mentioned above.

10.1.2 Compressors

These can be for air as well as gas compression application and can be in various sizes and categories.

Mainly three broad types of compressors are in use:

(i) Reciprocating:

Vertical and Horizontal.

Single cylinder, two cylinders, three cylinders and even four cylinders (4 stage) for Low pressure stage, Intermediate pressure stage and High pressure stage with inter-coolers and after-coolers having cold water circulated in the cylinder jackets for cooling.

(ii) Rotary:

Single stage, double stage, (Radial/axial) multiple stage, with intercooler and after coolers.

(iii) Centrifugal:

Radial or radial-axial flow configuration can be used depending on the requirement of pressure and discharge rates.

For compressed air production, both reciprocating and rotary types are used ***(NOTE: Rotary compressors are costlier but more efficient; they can be smaller in size for the same discharge capacity and consume less energy).***

For gas compression, generally high discharge, high pressure multi-stage rotary compressors/turbine type compressors are used (as in petro-chemical industries; as boosters in long distance gas pipe line etc.).

10.1.3 Power Presses

(A) Power Presses are generally used in industries for the following purposes:

   (i) Forming, bending, pressing of stacks.

   (ii) Straightening of metal bars, rods, plates etc.

   (iii) Forging and shaping of metal and non-metal compounds/work-pieces.

   (iv) Trimming, shearing of metal/non metal work-pieces;

(B) Power Presses come broadly in the following three categories:

   (1) **Hydraulically operated presses:**

   These can be in a wide range of capacities—say from 1-2 tonne to even Up to 10000-12000 Tonne (single cylinder to multiple cylinder configurations, depending on the pressing power requirements).

   Hydraulic presses are used for shaping and forming: bending and stack consolidating, forging of bigger work-pieces, bending of bars, plates, beams etc.

   (2) **Pneumatic Presses**:

   Normally in lower pressing force capacity ranges (operated by compressed air). Pneumatic presses can be operated by compressed air supplied through hose, and worked through the action of a piston inside a cylinder (can be reciprocating or non-reciprocating type).

The system draws compressed (normally 7kg per cm^2) from the compressed air pipe lines laid out in the shops;

(3) **Power Presses (Crank-operated and Screw type):**

Can be in wide range of capacities; crank type presses can be with capacity range from a few tonne to even up to 6000-8000 tonne.

Power presses are largely used for sheet metal works (forming, trimming etc.), by using relevant press dies.

Large capacity ones can be used for forging engine crank-shafts and cam shafts.

(C) Powering System for Presses:

(i) **Hydraulic Presses** get their pressing power from electrically driven hydraulic pumps (using water, high pressure hydraulic fluids etc.) connected by pipe lines through hydraulic cylinders which can operate individually or in tandem or simultaneously.

(ii) **Power Presses** are directly driven by electrical motors (or petrol/ diesel engines, or by steam engines) coupled to the crank shaft through electro -pneumatic clutches.

(iii) **Pneumatic Presses** can be operated by compressed air supplied through hoses, and worked through the action of a piston inside a cylinder (can be reciprocating or non-reciprocating type).

10.1.4 Hammers (Pneumatic)

Drop Hammers and Counter-Blow Hammers:

Mostly used for Forging operations.

These hammers operate on pneumatic system (compressed air); employed for hammering work-pieces in small sizes requiring lesser tonnage of hammering; Bigger size hammers with higher force and weights are used in forging operation on large work-pieces in a Foundry Forge Plant (**both drop and counter blow types**).

10.1.5 Furnaces and Ovens

A variety of furnaces and ovens are used in industries (including Foundry & Forge Plants, Process industries). We shall discuss the following commonly used ones:

(1) Furnaces:

* Furnaces are used in engineering industries for heating, heat-treatment, and in some cases for melting of metals, mostly steels.

*Heat Treatment Furnaces are used for hardening, tempering, annealing, stress-relieving, baking.

*Melting Furnaces are used for melting/re-melting of metals; in some cases, certain non-metals like plastics, resins, epoxy compounds also;

(a) **Types of Furnaces:**

The following types are generally used in Engineering Industries depending on the purpose and the material:

(i) Oil Fired Furnaces (furnace oil fired) (temperature range up to 1300°C)

(ii) Gas (Producer Gas/Coal gas, Natural gas, LPG) fired furnaces; (temperature range up to 1300°C);

(iii) Electric Resistances Heating Furnaces (temperature range up to 1100°C);

**(iv) Electric Arc Furnaces. (temperature up to 1550°C)

(v) Electric Induction Furnaces. (temperature up to 1550°C)

*(vi) Blast Furnaces (for producing pig iron from iron ores) (temperature 1450°C).

*(vii) Open Hearth Furnaces, for steel making: temperature-around 1500°C.

NOTES: *(i) These furnaces are used in Integrated Steel Plants.

*****(ii) These are used in Steel Foundries, but smaller versions are also used in heat treatment of metals (Salt bath furnace).***

(b) **Industrial furnaces are also classified as under depending on constructional features and application:**

(a) Chamber Type Furnaces,

(b) Bogie Hearth Furnaces,

(c) Bell Furnaces,

(d) Rotary Hearth Furnaces,

(e) Tube Furnaces,

(2) **Ovens:**

*Ovens are available in different sizes with different temperature ranges and are used for drying, baking, induced thermal reactions, curing etc.

*These are available in industrial models as well as in laboratory models.

*These are either electrically heated or heated with LPG burners, coal gas or kerosene-gas burners.

*Sizes may also widely vary from several cubic metres in volume, to less than a cubic metre in volume.

*Temperature range may vary from 100°C to 500°C depending on application and the constructional features.

10.1.6 Boilers

REMARK: Although there can be boilers for liquids other than water, we are limiting our discussion here to the boilers utilizing water for supply of hot water and/or for generation of steam for the purposes of:

(a) Process-Steam/hot water, (for processes or for human comfort use).

(b) Heating (district-heating in cold-weather condition in residential and industrial/commercial areas),

(c) Supply of steam for driving prime-movers like steam engines and steam-turbine for conversion of heat energy to mechanical power (rotary, torque), for driving electrical generator for generation of electrical power.

>>Boilers are made available in two basic categories from the principle of utilization of heat energy:

*** Unfired Boilers.**

*** Fired Boilers.**

(1) Unfired Boilers:

These are broadly of two types :

(i) **Heat -Exchanger** type (transfer heat from one medium to the other-through indirect contacts)- utilizing hot flue gasses/hot air or hot liquids/fluids from other working systems where heat gets generated as a bye-product/surplus.

>**Heat Exchanger Types** can be small, medium as well as large ones depending on the quantity of heat input available, volume of heating media, temperature and pressure of steam required.

> **Large Heat Exchangers (HRSGs)** are used for large-scale generation of steam for running turbines in Nuclear as well as in Combined Cycle Power Stations;

> **Large Heat-Exchangers** are also used in process industries like petro-chemical complexes, fertilizer plants, oil refineries, primarily for removal of heat generated as a by-product elsewhere in the plant, or for secondary use of surplus heat-transfer process.

(ii) **Electrical Resistance Heating Type**

These do not use any conventional fuels; these generally utilize electrical energy for heating by passing electrical current through heating elements (resistance heating). The electrically heated boilers are generally in small sizes, – for use in laboratories, hospitals, hotels, domestic purposes (for hot water or steam).

(2) Fired Boilers:

These boilers incorporate in them, a furnace or a fire-chamber where fuels such as Furnace oil, Natural gas, coal, lignite, Pit, or even wood or municipal garbage are burnt to produce enough heat to convert water into steam/super-heated steam.(*)

NOTE: Fired Boilers can also be used for boiling other liquids for certain process industries such as food processing, pharmaceuticals, paints.

(2-i) **Fired Boilers are of two basic classes: Flue Tube and Water Tube.**

(A) Flue –Tube Boilers:

In this type of boilers, hot gases emanating from the fire-chamber are passed through seamless steel/alloy-steel tubes-surrounded by water in a sealed water pressure vessel where heat from the hot flue gases gets transferred to the water which eventually gets converted into steam. Locomotive Boiler is an example, being used on steam locomotive engines. There is a industrial version of this type, fixed to the ground, and used for the supply of process-steam in industrial plants. There is also packaged-type flue-tube boiler used for supply of process steam. Flue tube boilers are also deployed for generation of steam for heating purposes in industrial plants; also used in cold countries for district heating in residential areas and offices by circulating steam through insulated pipe-lines which again passes through the radiators installed in rooms/halls.

Generally, flue-tube boilers are low capacity, low pressure and with lower range of superheat and therefore, are not suitable for large-scale mechanical and electrical power generation; they can, however, be used for driving steam-engines which act as prime-movers for industrial drives and for limited electrical power generation. Small size steam turbines-generator sets can also be run by steam generated in locomotive type or vertical type flue-tube boilers.

(B) Water-Tube Boilers:

In this type of boiler, water is passed through seamless metallic tubes arranged in large numbers around a fire-chamber where hot flue gases get generated (due to burning of fuels: oil, gas, coal etc) and get circulated around the water-tubes, and the water eventually gets converted into steam; the steam accumulates in the steam drums*—both low pressure and high pressure. In this type of boilers, large scale generation of steam (30 to over 3000 tonne per hour) at high pressure (35 to 300 ATA) with a high degree of super-heat (240°C to 650°C) is possible for driving steam turbines to rotate large generators to generate electrical power. In fact water-tube boilers can be designed and configured for supply of steam for power generation for running large steam turbines of even up to 1000 MW unit capacity.

NOTE: As per the latest technological trends, for large scale generation of steam for operating large steam turbines for power generation, 'Once-Through' type drum-less boilers are being manufactured by reputed boiler manufactures for better operating efficiency and better heat-rate.** ***(NOTE: Elimination of boiler drums in this type of boiler not only reduces the weight/material of construction almost by 25 to 30%, but also drastically reduces steam path lengths, reduces heat losses and considerably improves over all thermal efficiency vis-à-vis fuel consumption); **however, such high technology boilers operate in Super Critical Performance Parameters which not only necessitate the use of high grade/quality materials for the heat cycle structures, piping, headers etc, but also the use of high grade and reliable computerized controls and instrumentation systems, besides use of high grade fuels with high calorific values.**

Small size water-tube boilers are also available in **packaged-version** (mostly back pressure) for supply of process steam, and in some cases, also for supplying steam to small capacity steam-turbine for running small size electric power generators.

(C) Packaged Boilers:

These are smaller compact type boilers deployed for supply of steam at a rather low parameters for use in process plants(steam-heating, painting lines, sterilization, steam-washing). These are not normally used for power generation or for supplying steam for powering drives. These boilers can be portable also and can come in both water-tube or flue tube versions.

(D) Vertical Boilers:

These are similar to locomotive boilers with vertical construction, suitable for installation in locations where space is a limitation like in small factory premises, in steam ships. Steam parameters can be around: 10-50 T/hr, 200^{0}C to 400/440^{0}C, 33 to 60 ATA pressure.

In large steam ships, more than one such boilers can be installed depending on the capacity of the **marine turbines** installed therein.

(E) Fluidized-Bed Boilers:

These are a special category of water tube boilers designed to use low grade fuels (having lower heat rate in calorific value, particularly low grade high ash-content coal/pit).

The industrial and power-generating versions called **'Circulating Fluidized-Bed (CFB) Boilers'** incorporate specially designed mobile/circulating(rotating) fuel beds in the combustion chambers for optimal burning of and generation of heat from the burnt fuel, thereby increasing the over-all thermal efficiency of the system to a practically achievable optimal level. These boilers have unique advantage of low-emission of toxic gasses like CO,CO_2, NO_2, SO_2 etc.

Because of the special advantages and higher thermal efficiency, the demands for such boilers, both for process steam as well as for steam generation for power-plants, is increasing. Manufacturers (which includes BHEL in India) are engaged in developing higher capacity versions for using low grade coals to derive maximum possible heat-rates/heat-energy extraction in a cost-effective technology, for application both for power generation as well as for industrial process steam.

10.1.7 Material Handling Equipment

A. In an industrial complex, a variety of material handling equipment are deployed; **most commonly used ones are:**

(i) **EOT-cranes, Gantry cranes, Tower cranes, Suspension cranes, Cantilever cranes, Self-standing Jib cranes;**

(ii) **Inter-Bay and Intra-Bay Transfer Trolleys, Hydraulic lifting** Platforms, Scissor-lift devices, Pick-and-Place Robots, Loading-Unloading devices (pneumatic, electro-pneumatic, hydraulic, Electro-hydraulic, Diesel-electric etc.).

B. While the design and construction of the above equipment will depend on the magnitude of the load and the kind of application, some of these or a combination of many of these are deployed in work-shops as well as in project construction sites.

C. EOT cranes, Gantry-cranes and Tower cranes can have the lifting capacities ranging from 2 Tonnes to over 500 Tonnes, depending on the site of deployment and the magnitude of the load to be handled.

REMARK: Some special purpose heavy Gantry Cranes are also mounted on large Barges at Ports, and also deployed for salvage operations for sunken vessels/ heavy equipment; these are NOT used in industrial operations.

D. While some of the above handling equipment can be powered by hydraulic systems or by diesel engines, or by electrical motors (backed up by battery bank for emergency operations), most of the higher capacity ones (say above 2-3 tonne) are powered by electricity from the mains to drive suitable capacity electrical motors with matching control-gears and trailing (power supply) cables wherever required.

10.1.8 Transportation Equipment:

These include:

(i) **Motor Vehicles** of all types,

(ii) **Railway(Locomotive) Engines and Wagons** (there are a variety of types of railway wagons, depending on the use and the carrying capacity),

>> Larger industrial units, mining complexes and power stations use Diesel Engine driven, or Diesel-Electric Shunters in their premises/yards.

(iii) Transport Air Crafts:

These are not normally deployed in industry; as and when air lifting is required, such services are hired from the air transport agencies.

(iv) Inter-Bay and Intra-Bay Trolleys:

These are driven by diesel engines or by electrical motors coupled to winches, or with on-board geared electric motors; these trolleys have flat top platforms and are used in factory blocks/workshops for internal material movements on shop-floors; electric motors draw power from underground trolley busbars through current collections.

(v) Battery Operated Platform Trucks;

These are deployed in the work-areas for internal as well as short-distance movement of materials weighing up to 2-3 tonnes; normally these are powered by DC Motors drawing power from heavy-duty batteries (36 or 48 Volts, arranged in banks).

(vi) **Air Castors:**

Compressed Air inflated cushion-mats floating trolleys moving on the floors, can transport loads (and 100 kgs to 1500 kgs) around/across the shop floors by exerting little push-efforts by a man; can be very convenient in the assembly-lines for items like control panels, refrigerators, control desks, switch-gears, copier machines and the like.

10.1.9 Earth Moving Equipment

The following are **commonly deployed** earth moving equipment in Industrial Complexes:

(i) Bull Dozers (ii) Front-end Loaders

(iii) Rear Dumpers (iv) Excavators ,(v) Shovels.

The above mobile machines are mostly powered by diesel engines except excavators which can have diesel-electric combination drive system.

10.1.10 Mechanized Ventilation Systems

A. Mechanized Ventilation Systems are deployed where natural ventilation is not adequate, and forced ventilation becomes imperative for:

(i) Proper circulation of fresh air/clean air in the work area,

(ii) Removal of hot air, fumes, gases, dusts etc. generated due to activities/equipment operating in the work-shops, Labs/Kitchens, process plants etc.

B. Forced (Mechanized) Ventilation System generally comprise of fans/blowers and connected ducting systems, with matching drives and drive controls for the fans and blowers. Structural supports for ,as well as proper covering of the ducts for aesthetics/insulation/protection, are suitably placed; baffles in the ducting system are also to be necessarily provided for controlling air/gas flows through the ducts..

C. Ventilation system also becomes an essential sub-system for any Air-conditioning (AC) system, but in this case, the ventilation equipment and the associated ducting etc. become a part of the total AC Plant and equipment installations.

In Storage-Godowns, workshops, laboratories etc., roof mounted ventilators with motorized fans are also installed.

D. For the Air Conditioning System, when make-up fresh air is taken in from outside, to filter out the dust-particles, specially designed **air filtration units** are installed and the filtered air is passed on to the Air Handling Unit (AHU).

Commonly, Dry Type Filters are used.. In a very large Air Conditioning System or in an AC installation where very stringent clean room facility is also required, Electrostatic Filtration System may be adopted.

10.1.11 Fans and Blowers

(A) Fans and blowers are used in work shops, labs, offices and stores for:

- For providing cooling (employees, equipment/working systems).
- For ventilation : circulation of air, creating draught for driving out fumes, stagnant hot air, toxic gases, flue gases from furnaces, boilers; from labs, offices, work-shops etc.
- For blowing of gases and/or air into pipelines, into furnaces/boilers and into other relevant working systems including Summer Air conditioning ducting system/Air-handling Unit (AHU).
- For creating forced draughts in the boiler furnace/fire box and evacuating flue gas into the stack system.
- For blowing fresh air/oxygen into boiler fire boxes and furnaces for facilitating efficient fuel burning process.
- For gas transmission though a long-distance pipelines including those for flow booster-operations.

(B) As already indicated above, these fans and blowers are driven by electric motors of matching types and capacities (hp/Kw) along with matching drive controls and switch gears.

(C) The sizing, capacity and types of Fans and Blowers will depend on:

(a) Volume of air/gas to be handled,

(b) Characteristics/quality of the air/gas/fumes,

(c) Constraints related to location,

(d) Length and size of the ducting system,
(e) Stack system and size for forced evacuation of flue gases (like that in a thermal power station).

10.1.12 Air Conditioning Systems

(A) Winter Air Conditioning: (heating/warning up spaces, together with humidity control):

In places where ambient temperature goes below say 20^{0}C (in some cold weather places, the ambient temperature may go even below (–) 20°C), adoption of Winter Air Conditioning System becomes imperative (with matching RH control), both for human comfort as well as for process-control; also to avoid detrimental effects of low temperature on certain materials.

The Winter Air Conditioning can be done by adopting any or a combination of the following systems:

(1) **Fire-Place in homes, offices:**
In homes and offices, a Fire place is located in a suitable location (with proper Chimney System) for burning wood, coal, gas etc. to heat up the space. In such a case, at least one entry point for ingress of fresh air has to be provided for supply of O_2 to the burning point.

(2) **Using Electrical Resistance Heaters to heat up/warm up the desired spaces/areas:**
(i) For a small confined space, ordinary electric heaters with parabolic reflectors may be enough (heating through radiation.)
(ii) For comparatively bigger rooms/spaces, electrical resistance heaters associated with a blower can be used for circulating hot/warm air around.
(iii) For a bigger room/hall/laboratory/work-shop, multiple electrical hot air blowers may have to be deployed and placed at suitable locations.
(iv) In some models of Window type AC Units, electrical heaters are also incorporated and by using the fan/blower of the Unit, hot air circulation can be done. During summer, such electrical heaters are cut off from the system.
(v) **For large halls/premises**, a Centralized Electrical Heating System with a bank of electrical heaters and matching Air Handling Unit, fans and blowers can be installed in a suitable location and the Air handling Unit can blow hot/warm air into the existing ducting and exhaust grill system (meant for

Summer AC) thereby discharging and circulating hot/warm air into the targeted premised (of course with due control on temperature rise and humidity through appropriate associated control devices).

NOTE 1: In such a case, the ducting system will have to be effectively insulated from the summer-air conditioning plant (comprising of gas/refrigerant Compressors, their drives and controls, chilling water coils etc.); temperature of heated air going through the ducting system should be within the maximum range which is safe for the ducting system.

NOTE 2: Electrical space heating can also be done by using high-wattage electrical incandescent lamps/infrared lamps arranged suitably in a desired place, but in this method, only a small volumetric space (say around 60 to 100 cum) can be conveniently heated, but this is mainly used for industrial heating, drying, curing, baking processes.

(3) **Solar Heating Option**:

There is also a solar heating system where water is directly circulated through the solar parabolic reflector panels by means of suitable pipelines (with thermal insulation covering the part of pipes outside the reflectors). Hot water thus generated is passed through a suitable insulated pipe lines laid out in the desired places. These are generally placed on the roof-tops of buildings.

This system avoids electricity consumptions and associated costs.

(4). Hot-Air Circulation: Air heated by hot flue gases emanating from Combustion Chambers/Furnaces:

(i) By burning coal, furnace oil, fuel gas etc in a combustion-chamber or furnace, hot flue gases are generated. These flue gases can be passed through a heat-exchanger type of Air Handling Unit where air gets heated (by indirect heat transfer process) and this heated air is then blown through the existing AC ducting and grill system into the areas to be heated. Flue gases containing harmful emissions (NO_2, SO_2, CO, CO_2 etc.) are led out by a suitable separate exhaust system and discharged to the atmosphere, through suitable gas scrubbers/pollution control devices.

(ii) Wherever such combustion chamber or furnace is operating for any other industrial processes, the hot flue gases emanating from there can be conveniently used **(Waste Heat Recovery Process = WHRP)** for such heating of space by using hot-air blowing through suitable discharge system (grills).

(NOTES: (i) In this case also, if existing AC ducting system is used for circulating hot air, the AC plants and the associated chilling plant, controls etc. have to be isolated from the ducting system).

(iii) Additionally, in such a system of heat generation by using combustion of coal, fuel oil, fuel gas etc., adequate measures must be taken to ensure proper handling and storage of fuel, environmental/ecological and safety aspects, (incoming delivery/ receiving points, storage tanks, dispensation pumps and pump-drives and controls; fire-prevention, fire-fighting system, servicing, maintenance etc.).

NOTE: ***Such a system should be least preferred because of associated environmental and ecological problems. If at all adopted, the heating plant may have to be installed in a separate place away from the premises to be heated; hot-air can be circulated through blowing, through insulated ducting and grill system.***

(5) District Heating in Cold Climate Countries (where the ambient temperature goes much below zero (sometimes even below –10°C), the comfort level deteriorate for a considerable period in winter months):

A Centralized Fired Boiler Stations using coal/furnace oil/ natural gas as Fuel, generate low parameter steam which is supplied to the residential areas and offices through insulated pipelines connected to the radiators placed in the user premises; this is a costly option and also have associated environmental problems because of harmful emissions emanating from such boiler stations, needing additional capital expenditure towards installation of pollution control devices.

Because of many advantages, unitary electrical heating devices are preferred.

B. Summer (Hot Climate) Air Conditioning:

Depending on the premises (size/volume, location) to be air-conditioned, temperature and humidity (RH) and other ambient conditions desired, **any of the following methods are generally adopted:**

(1) Window type AC Units:

These can be used for comforts of personnel, or for use on machines and equipment, or for material storage. These can be installed when the total space-volume of the premise is small (say up to 300-400 Cum) and the temperature and RH required are in the range of 20°C-25°C and 60-70% respectively.

There are **standard models** in the ratings range of 0.5T, 0.75T, 1-T ,1.5T and in some cases 2-T per hour (rate of refrigerant circulation) are available in the market.

There are also the advanced technology versions known as **'Split-type' and 'Triomatic-Type'** models in the rating rage of 1.5-3T/hr. are available for installation in premises where low noise-level is desired and where more efficient cooling in a compact area becomes desirable.

Specially designed Compact units, using special class refrigerating fluids for effective in-situ cooling on certain machines and machine-control panels (like CNC machine tools, process lab equipment) are also available.

(2) Packaged Type AC Units:

When larger space (halls, bigger stores, medium size auditorium etc.) (say from 500 CUM to 3000 CUM ranges) is to be air conditioned for an ambient temperature of around 20°C (20°C to 25°C) with RH at 55% to 65%, packaged type AC Units can be installed. Since these are of compact design and construction, the equipment require a smaller space for accommodation. The refrigerant compressors for these packaged units can be in the capacity range of 5 tonne/hr. to 15 tonne/hr. They also require a small cooling tower (for cooling circulating water for compressors and AHU) which can be installed outdoors.

(3) Central Air Conditioning Plant (with associated chilled water plant and cooling tower):

When the volume of space to be cooled is large, say above 3000 Cum, for efficient and dependable Air Conditioning, larger size Centralized Air-Condition Plant, with associated chilled water plant, chilled water pipe lines and AHUs, is required to achieve the desired temperature (around 20°C) and Humidity (RH:55% to 65%). The capacity range of individual plant may very form 20T/hr. to 120T/hr. and multiple units can be installed in a Central place (in the basement of a building, or in a separate adjacent building) depending on the size of the areas to be cooled (AC load). The chilled water (at around 7°C at the entry point) in insulated pipelines may be taken from there to various locations/floors in the building and connected to AHUs installed there. Cooled Air with permissible level of RH values from the AHUs are blown into the ducting system going to various rooms, halls, zones and gets discharged through the grill system. The AHUs are provided with air filtration units (Dry type or electrostatic type) to make the air dust-free to the prescribed level. In such a system, at least one plant is kept as a stand-by unit. The chilled water pipe lines and the ducting are insulated from external/atmospheric heat. In AHU, arrangement of replenishment of fresh make-up air from outside (around 15%) with filters is also incorporated.

10.1.13 Refrigeration

(i) While the Refrigeration Process is also associated with Air Conditioning (chilling plants), in some cases for storage of some types of materials/perishable items at lower temperature, (say below 15°C to even at (-) 15°C), refrigeration plants of proper capacity are installed (commonly known as Cold Storage). Besides the household refrigerators which can achieve a low temperature of around (–) 7°C in its freezer chamber, there are commercial/industrial models of refrigeration system including deep freezer units which can achieve temperature as low as (–)15°C.

(ii) **There are special types of Deep Freezers** used for producing Liquefied Nitrogen, and Liquid Helium at temperatures as low as (–) 196°C(77°K) and (–) 269°C (4°k) respectively, required for industrial and R&D applications. **Very low temperature refrigeration process is also known as 'Cryogenic Process'.**

(iii) While the normal category of refrigeration processes including cold storages and ice-making plants use Freon-12, Freon-22, Freon-502, Brine Solutions, Ammonia etc. as the convenient refrigerants, for the deep freezing and Cryogenic processes may use CO_2 crystals (Dry-ice), propylene glycol, Liquid Nitrogen, Liquid Helium, Lithium-Bromine solution as the refrigerants.

(iv) **The general principle of refrigeration process** followed for the above Systems, is to compress the refrigerant to a high pressure and then pass on the compressed fluid to a condenser where pressure drops suddenly/quickly and quick evaporation **(vapour absorption technology)** takes places (due to low boiling point at low pressure/ vacuum condition), resulting in release of heat to the surrounding atmosphere. Repetition of the process results in a cumulative effect for removal of heat to the desired level. Water or other fluid to be chilled/ iced or liquefied is circulated around the evaporator coils (carrying the refrigerants) and in turn gets cooled/refrigerated to the desired level. To hasten the cooling and evaporation, a cooling tower is also associated for taking away heat from the condenser/evaporator coils and discharge the generated/extracted heat to the atmosphere.

NOTE: The above paragraphs aim at presenting a brief recapitulation on the subject. For more in-depth knowledge, readers are advised to read text books on the subject. This book does not propose to go into theoretical aspects of engineering/design and construction aspects of these systems.

10.1.14 Humidity Control

Humidity Control (Relative Humidity = RH) assumes a very important role in the proper working in certain locations and under certain circumstances. While human comfort prefers a RH value range between 45% to 65%, this range for certain materials, equipment, specially sensitive electronic components/ gadgets, may go much lower, around 35% to 55% (maximum), since presence of moisture/water vapour beyond the prescribed limits is very detrimental to such materials, equipment/components, and may reduce their working life drastically, besides adversely effecting the working of the related system.

On the other-hand, too low a RH value, say below 30% may also have detrimental effects on life/durability of certain materials like wood, concrete work, textile fibres, paints and varnishes etc, besides creating discomfort for human beings (sore/parched throat, dryness/irritation in the mucous membrane inside the nose).

For Humidity Control, gadgets/equipment for both increasing the humidity level, or decreasing the humidity level, may be required to be installed depending on the ambient conditions desired at the site.

(A) **Humidifiers:**

When RH value falls much below the desired level, Humidifier is brought into operation. The equipment adds water vapour into ambient space, or into the flowing air in the existing ducting system. The rate of water vapour injection can be controlled by suitable devices incorporated in the Humidifier. Water is drawn from the water supply pipeline and vaporized; the process is controlled by a device called **'Humidistat'** which automatically cuts of water supply as soon as RH value goes up beyond the permissible/preset value; the device restarts as soon as RH value goes down below the minimum permissible value. Several types of Humidifiers are available being marketed by reputed manufacturers; these may be any of the following types:

* (i) Pneumatic Humidifier.
* (ii) Evaporative Humidifier.
* (iii) Steam-jet Humidifier.

Selection of any of the above will depend on the application and the location of use/site conditions.

(*NOTE: For a more in depth knowledge on the subject, readers are advised to read test books on the topic.)

(B) **De-Humidifiers:** When RH level goes up beyond the desired/ permissible level, use of De-humidifier becomes imperative.

This can be achieved in the following manner:

(i) Installing and activating devices containing chemical dryers such as activated carbon or silica gel.

(ii) Installing electrical heating device (resistance heaters, high wattage electric incandescent lamps, infrared lamps) and thus driving out moisture/water vapour, but this can be done only on a limited scale and for smaller confined spaces.

(iii) Chilling the humid air below its dew points, thereby removing the excess moisture (Condensed in the form of water/or ice), followed by reheating the de-moisturized air space to restore the desired level of temperature and released to the space around.

(iv) Force-ventilating (by using exhaust fans/blowers) in some confined areas like kitchen and laundry operations and the like (Stores, bathrooms, attic spaces etc.).

(v) Damp proofing land areas under the building structures (by using anti damp compounds, special moisture proof coating etc.).

Standard models of De-humidifier are available in the market employing any or combination of the above mentioned working principles.

10.1.15 Piping Systems

Industrial complexes may have a variety of indoor and outdoor Pipe Lines carrying steam, water, oil, hydraulic fluids, gas, chemicals, slurries, sewage etc., in some cases, at high temperature and pressure. Some of the pipe lines will have special thermal insulation cover on them to prevent loss of heat (as in hot water/steam supply pipes).

While the indoor piping may be laid along the floor, or walls or the ceilings or on special piping support structures, with suitable fixtures/brackets, the outdoor piping may be laid over ground on concrete or steel structure piers, or even in trenches in the ground, depending on the types and usage.

Depending on the material being carried, the pipe lines are colour-coded for proper identification to avoid confusion and wrong inter-connection. While deciding the piping layout and sizing the same (as per design procedures and as per the system requirements), care has to be taken against structural failures (of fixtures/supports), mechanical damages, failures of flanged joints and the suitability and quality of the piping materials and valves being used. **National and International standards as applicable (API/IS etc.) must be followed to avoid substandard installation and chances of failures/mishaps.**

10.1.16 Incinerators (For Waste Disposal)

Need for these devices often arises in industrial complexes, hospitals and in Municipalities for disposal of wastes/garbages. There are a large variety of them available in the market. Depending on the application and the type of wastes, these can be selected and installed at a suitable place, but preferably away from residential/populated areas as they may, during their operation, give out obnoxious emissions (gases, dusts, smokes, fumes etc.).

Selection of appropriate type of incinerators must precede a thorough knowledge of the types of wastes to be processed. The American Institute of Incinerators(AII) has classified the Wastes into six types (Type-1 waste to Type-VI waste) and the Incinerators into nine broad classes: Class-I, Class-IA, Class-II, Class-IIA, Class-III, Class-IV, Class-V, Class-VI, Class-VII; (for further details, refer to Handbook of AII).

Besides the above, the Trash Burners having a maximum volumetric capacity up to 12 cubic-feet (cft) has been considered as domestic equipment and beyond this size, as industrial type attracting pollution/environmental regulations in their design and operating methods. As regards the material of construction of the incinerators, safety requirements etc. reference may be made to the relevant standards of the (US) National Fire Protection Association, or any other available National or International standards. Local Government Pollution Control Agencies, Municipal Corporation regulations governing such incineration must be followed.

10.1.17 Hydraulic-Powered Operating Systems

(A) Hydraulically Powered Devices/Equipment are one of the most prevalent operating systems incorporated in many Mechanical Plant and Machinery/Mechanical Working Systems and uses water or hydraulic fluids/oils as the medium of transfer of power. The systems employ specially designed pumps, hoses/pipes, cylinders, pistons/rams, levers and linkage mechanisms besides the pump-drives (engines or electrical motors), drive controls, flow-control devices (such as valves, actuators) and instrumentation (pressure gauge, temperature gauge, flow meters etc.). Such hydraulic devices mostly work at room or near room temperature regimes (without any appreciable temperature rise) and transfer power smoothly unless specially intended and designed to act quickly/instantaneously in certain applications by the use of quick-acting actuators and rams.

(B) Depending on the pressure range, magnitude of power transfer involved, and the quantity/volume of hydraulic fluid to be handled, the hydraulically operating systems may incorporate pressure accumulators for reliable and smooth working. Coupled with electro-mechanical or electro-pneumatic and/ or electronic systems (electronics mainly in controls and instrumentation). The hydraulically operating systems find wide applications in industrial plants, machineries (machine tools), transportation equipment (automobiles, railway wagons/ bogies/engines), aircrafts, material handling equipment, mining and earth moving machineries, automatic process plants and many other mechanical devices.

(C) One of the most important advantages of hydraulically operated systems is the smooth transfer of power and movements without resorting to complicated and avoidable mechanical linkage and geared mechanisms, thereby reducing maintenance efforts, power transmission losses, and facilitating easier maintainability of the operating systems besides reducing costs towards installation and maintenances.

(D) Based on the equipment design and the application, the manufacturers recommend the type and grades of hydraulic fluids to be used and the user should follow the recommendations as best as possible and practicable.

(E) Depending on the pressure ranges required, the volume of fluid to be handled, and the range of power transfer, the hydraulic pumps used can be any or a combination of the following types:

(i) Screw Pumps/Rotary Gear-Pumps;
(ii) Rotary Vane Pumps;
(iii) Reciprocating Pumps (with Pressure Accumulators);
(iv) Positive Displacement Axial Flow Multi-Stage Pumps;
(v) Turbine Pumps (Mainly Axial Flow Type);

(F) Another important area of application of Hydraulic Power is in the form of different types of Hydraulic (water) Turbines (Pelton Wheel, Francis, Kaplan etc) used as prime-movers, – particularly in Hydro-Electric Power Generation Stations.

10.1.18 Pneumatic Systems

(A) Pneumatic Devices and Systems using compressed air is a very common working system in most work-shops, and in many machine tools and equipment.

The most prevalent systems have a working pressure between 100 Psi to 120 Psi (around 7 to 8 kg/cm^2).

(B) The Compressed Air supply is used commonly for:

(i) Operation of pneumatic hand tools like nut-tightener, buffing and hand held grinding machines, pneumatic hammer (for concrete cutting/dismantling) etc.
(ii) Forge Shop Hammers, foundry equipment like vibratory moulding tables etc.

(C) For Machine Tools and other mechanical equipment employing pneumatic working system:

(a) For cleaning (blowing away dust, chips etc.)
(b) For operating pneumatically/electro-pneumatically operated actuators, flow control valves, pressure-switches etc.

(c) For pneumatic cylinders for pressing plates, small size rods/ bars;

(d) For operating **pneumatic clutches** on crank/power presses, pneumatic rams etc.

(D) Compressed Air at the desired pressure is obtained by operating Air Compressors of different designs and capacity-depending on the volume of air and pressure to be handled. ***(NOTE: Please refer to para -10.1.2 above about compressors).***

(E) The Compressed Air is normally distributed through pipe lines to various work areas. These pipe lines are connected to the mains coming out of air-receiver (accumulator) installed adjacent to the Air Compressor House (these compressed air receivers/accumulators store compressed air up to certain volumetric capacity at the rated system pressure and serves to supply compressed air at uniform pressure to the user points through pipes*.

(F) In some special cases of application, compressed air supply from the distribution pipe-line network is not very suitable due to pressure drop at a long distance; therefore, it becomes necessary to install smaller compressed-air units (with air receivers) near the load center ***(NOTE: Such units can be portable sets also with electrical motor or diesel engines as the prime-mover).***

***Remark:**

Pipes and pipe fittings use must conform to API or IS standards. The same is also applicable to hoses if used.

10.1.19 Oxygen Gas (O_2) Generating Plant

(1) When the requirement of O_2 gas is substantial in large engineering industry, the organization prefers to install it's own O_2 generating plant in the premises. The plant draws air from the atmosphere, separates O_2 and N_2 by using the selected technological process and stores both gases in separate receiver tanks from where the same are distributed through dedicated pipe lines(duly colour-coded) to the consuming points.

(2) **The plant comprise of three sub-systems: Air Compressor Unit, Air Separation Unit, and Storage/Receiver Tanks;**

(3) There are associated/dedicated piping and plumbing systems, Pumps and associated drives and controls.

(4) There can be smaller amount of CO_2 gas as bi-product, normally released to the atmosphere unless this find use elsewhere.

10.1.20 Acetylene Gas Generating Plant

(1) Some large and medium industrial units may require substantial quantum of Acetylene Gas for the purpose of gas welding and metal-cutting, and therefore, to ensure adequate and reliable supplies, may install their own captive Acetylene Plant in their premises.

(2) The plant comprises of Reactor Chamber inside which measured quantity of lumps of Calcium Carbide (specified sizes) are sprayed with water in a controlled manner; the resulting chemical reactions liberate Acetylene Gas which is pumped into storage cylinders ; cylinders in prescribed sizes are supplied to the consuming points. Since Acetylene gas is highly combustible/explosive, supply through pipe lines are avoided.

(3) Acetylene Gas plant is located in a isolated place, with all precautions and Safety measures put in place to avoid/minimize mishaps.

10.1.21 Producer Gas Plant

(1) Producer Gas is basically is coal-gas derived from Steam/Slack coal by processing the same in a furnace like Reaction Chamber with rotating hearth on which coal is fed through chutes , subjected to semi-burning by an array oil/gas burners ,and over the this burning coal, water is sprayed in a controlled dosing; the water particles get converted in to steam which in turn reacts with the flue gas emanating from the rotating hearth and thus get converted to coal gas.

(2) The generated gas is combustible and are fed to furnaces by pipe lines.

(3) Since the producer gas plant is a continuous process plant, it can not be shut down at will; during the low (or no) consumption period, the gas generation process is put to the minimum regimen, but even at that low regimen, certain amount of surplus gas gets produced which cannot be stored and has to be burnt out , discharging the flue gas to the atmosphere.

(4) As bi-products, the plant produces coal tar/bitumen and liquid phenolic acid; arrangements have to be in place for their safe disposal in conformity with the prescribed pollution control regulations.

(5) The gas produced will have lots of coal particles and vaporized bituminous substance which are needed to be removed; this is done by passing the gas through the Electro-Static Precipitators (DC High Voltage Filters).

10.1.22 Other Mechanical Working Systems

Besides the mechanical systems mentioned above under paras 10.1.2 to 10.1.21, there can be other mechanical working systems mentioned below for the purpose of general-interest information.

(NOTE : many of these are normally not deployed in industrial units)

Although the **list can be a long one,** those interested can go into further details about the ones of their preferences by studying the literatures/books on the subject and also by interacting with the concerned agencies/experts:

(i) Road construction equipment (road roller, bulldozers, hot-mix plants etc.)

(ii) Foundry equipment of various kinds (including melting/re-melting, / smelting , forging, rolling/re-rolling, drawing etc.)

(iii) Ships, Steamers, Motor-boats, Hover-crafts, War-ships, Submarines etc., Ship-building/Ship repair/Servicing yards.

(iv) Equipment and structures for cement, sugar, fertilizer plants, paper and paper-pulp mills, coal gasification plants, coal-washeries, textile mills, agricultural machinery/implements, **mining machinery**.

(v) Equipment and structures for Oil Rigs (both off-shore ON-shore), oil refineries, petro-chemical and fertilizer plants, drug and pharmaceutical industries.

(vi) Equipment and structures for power generation plants of various kind.

(vii) Equipment and structures for Sea-Ports, Dock-Yards, Ocean mining.

(viii) Equipment and structures for steel mills, other metallurgical plants, extrusion plants, power generation plants, nuclear reactors, petro-chemical plants.

(ix) Equipment for timber processing, food-processing plants, also plants for manufacture of drugs and pharmaceuticals, paints, varnishes, medical equipment.etc.

(x) Defence equipment like rifles, field guns, rockets, missiles, ammunitions, tanks, military vehicles; equipment and machinery for their manufacture and testing.

(xi) Aircraft and helicopters of various kinds, space-exploration vehicles and equipment etc.

(xii) Tea manufacturing equipment, other beverage manufacturing plants.

(xiii) Reprographic and Photographic equipment (Ammonia printing, plain paper copiers, cyclostyling machines, microfilming camera, microfilm processors, etc.).

(xiv) Different types of printing machinery.

(xv) Machinery for Mints, currency notes printing, stamps printing.

(xvi) Machinery for dairy products, cosmetics, telecommunication, computer system, tanneries,........ and the **LIST CAN BE ENDLESS**.

10.1.23 Energy and Power Supply System for Industrial

UNITS: Please refer to Chapter-12 for details on the topic.

10.1.24 Industrial X-Ray and Gamma Ray Equipment

(For in-process Quality Checks and Controls):

These are special purpose equipment for radiographic tests (Non Destructive Tests (NDT)) on structures, components and parts of industrial plants and machinery, mainly for ascertaining defect-free construction to ensure quality and reliability.

These are deep penetration devices which uses high intensity X-Rays and Gamma-Rays to investigate cracks, voids, inclusions, internal distortion etc. in castings, forgings, rolled stocks and welded parts to help making decision on:

- Acceptance or rejection,
- Or acceptance with deviations,
- Reworking for possible rectification for acceptance etc.

(A) **Industrial X-RAY equipment:**

(1) These operate through a specially designed extra-high voltage (100 Kv to 300 Kv AC ranges) transformer and a X-ray Projection Tube Head with associated controls.

(2) Industrial X-Ray machines are high intensity energy radiation emitting equipment and is therefore DANGEROUS to living beings unless provided with adequate biological shields; operators are supposed to strictly follow the established/ prescribed safe operation and exposure procedures.

Besides operating on dangerously high level of electrical potential (voltage), the X-ray beam itself can cause biological damage to living beings /personnel working around/nearby. Extra precautions are to be taken for safe operation.

(3) During Radiography, a part of the total X-ray beam is absorbed by the object and the cassette. Dangerous radiation will be present behind the object due (to penetration) and around the object in all directions due to scatter phenomenon. It is, therefore, essential to take such precautions that persons are prevented from entering/ coming near the X-ray beam projection zones.

Generally, it is recommended to **take the following precautions**:

- Direct the X-ray beam towards the solid concrete floor or wall

  The wall should be thick enough to stop ray penetration through the wall.

- Provide for a lead-lined (Pb-lined) screen (biological shielding)which absorbs all X-rays behind the object (As recommended by reputed X-ray equipment manufacturers):

For 130 KV	—	3 mm thick lead wall;
For 160 KV	—	4 mm thick lead wall;
For 200 KV	—	5 mm thick lead wall
For 300 KV	—	12 mm thick lead wall

- Restrict admittance to X-ray room to only the authorised and trained persons.
- To avoid dangers from scattered radiations, keep a distance of at least 6-10 metres from the unit during exposure (depending on the KV- range and X-ray intensity ratings).

NOTE: In the open area, individual X-ray beams may be considered dangerous at a distance of at least 100 metres from the source. Industrial X-ray should therefore be carried out in a separate enclosure with thick concentric concrete-lead-lined walls around.

(B) Industrial Gamma Ray Equipment:

(1) Like the Industrial X-ray machine, Gamma-ray equipment is also being used as a high penetrating radiographic equipment for NDT process in industry for detecting defects in parts and structures of plants and machinery (castings, forgings, welded parts, etc.) for cracks, voids, inclusions, internal distortion etc.

(2) While X-ray machine uses high voltage electric power for activating radiation source, the Gamma equipment uses **radio-active Isotopes like Cobalt-60, Iridium-192, (these two most commonly used), Thulium-170, Ytterbium-169 etc., as radiation sources.**

(3) Like the Industrial X-ray machine, the Gamma ray equipment, using radio-isotopes, have to be very carefully handled and operated as per prescribed precautions and procedures. In India, special permits/certifications is to be obtained from BARC (Bhabha Atomic Research Centre, Trombay) for procuring and using radio-isotopes and operating the Gamma ray equipment. **Any statutory regulations of the country are also to be complied with.**

(4) **Application:**

(i) Gamma ray process of radiographic tests are carried out for a job requiring a limited area of exposure but deep penetration on thick work piece which are difficult to handle and set inside the X-ray room; also for work pieces not requiring very stringent quality regimes.

(ii) Gamma-Ray equipment is portable and can be used on work-pieces in-situ condition.

10.1.25 Material (Metal and Non-Metal) Strength Testing Equipment

Commonly used machines are:

- Universal Testing Machines
- Tensile Testing Machines
- Sheer Strength Testing Machines.
- Tearing Strength Testing Machines.
- Hardness Testers: mostly manual with mechanized electro-mechanical and/or electronic measuring and indicating devices.

Most of the above machines are operated by electro-hydraulic power/drive system with associated measurement system/ instrumentation for recording of data (these employ electro-hydraulic/electrical/electronic instrumentation). The modern technologically advanced equipment is associated with microprocessors and/or PCs for control, measurements and recording.

10.2 PRIME-MOVERS: THE MOVING-FORCE DEVICES THAT DRIVE INDUSTRIAL AND TRANSPORT OPERATIONS

For industrial drives, power generation, transportation, mining, oil exploration, fluid pumping, material handling and also for domestic appliances, application of forces for linear, nonlinear, reciprocal and rotary motions are essentially required. To create such forces and cause movements, a category of machines/ devices have been developed and are being deployed for convenience of use; these are broadly called the **'Prime-Movers'**. These, either coupled directly or indirectly (through intermediate devices) to the gadgets/devices/machines/ equipment to be driven, **(since these driven systems do not normally have inbuilt system for effecting desired movements**). We shall discuss, in the following paragraphs, the most commonly used prime-movers. Most of these, by suitable arrangements of **kinematic linkages(application of kinematic pair concepts)**, can cause any kind of motion as mentioned above, with required capacity in KW/HP incorporated by design and construction:

(1) For Domestic Gadgets/Appliances

(a) Electrical Motors:

- Now a days most domestic appliances/gadgets operate with electric motors.
- Fractional Horse Power (FHP) Single phase AC motors of different types or in some cases universal (AC/DC) type motors are being used for driving household devices/gadgets like ceiling fans, table fans, small pumps, sewing machines, Air Conditioners. Mixer-grinders, Desert Coolers, Washing Machines, Refrigerators etc.

- 0.5 to 5 KW single phase AC motors are also used for domestic water pumping system depending on volume of water and the lift height required.

(b) Petrol Engines for small portable power generating sets are also in use for domestic or office emergency power supply. These petrol engines in compact air cooled designs, are available in ½ KW to 3KWs ranges.

(2) Prime-Movers for Industrial Applications:

(a) Industrial units (manufacturing as well as process) require various types of **driving prime-movers** for linear, non-linear, reciprocating and rotary motions for driving machinery and plants/equipment, vehicles, transportation, material handling equipment etc. Most predominantly used **prime-movers** for industrial drives are electrical motors, as they are the most convenient and energy efficient machines (efficiency can be as high as 98% depending on design and category).

(b) There can be other types of **prime-movers** like Petrol and Diesel engines, Steam turbines, Steam engines, gas turbines, hydraulic turbines etc., depending on the exact requirements with respect to needed power capacity and application.

(c) Electric Motors can be DC supply operated or 1-ϕ/3-ϕ AC supply operated at low voltage, medium voltage and high voltage; capacity wise, they can range from fractional horse-power to over 20000HP, depending on application and power (**torque delivery**) requirement for the driven devices/equipment **(Remark: some years ago, ABB has manufactured and commissioned a few 40000 KWs AC-Variable Frequency Drive (VFD) 3-ϕ AC Synchronous Motors in Norway (North Sea gas fields) for booster pumping system for undersea gas pipelines; this is the highest capacity electric motor so far constructed and commissioned for industrial use).**

(3) Prime-Movers For Power Generation Pants:

The prime-movers for power generation plants can be Petrol engines, Diesel engines, Steam Engines, Steam turbines, Gas turbines, Hydraulic turbines, Wind propelled turbines, depending on the type and capacity ranges of the power plant. **Please refer to Power Plants discussed in Chapter 12 for more details.**

(4) Prime-Movers For Transportation Application:

(a) For Railways : — Electric Motors in AC/DC (EMU)
— Diesel Engines
— Steam Locomotives

— Diesel- Electric Engines .

— Magnetic Levitation ($$) (Electro-Magnetic System based) for urban MRTs(*), even for long distance high speed rail system.

NOTE: Long distance Superfast Express train system (popularly Known as Bullet Trains) is now under operation in China, Japan, France on this technology. Recently, such a train system has been commissioned in China in Shanghai region (500 Kmph).

(*Mass Rapid Transportation System = MRTs).

(b) For Road Transport Machines/Vehicles

— Petrol and Diesel Engines;

— Battery operated Electric Motor for electric vehicles.

— LPG fueled Engines (mostly modified petrol/diesel engines).

— Fuel Cell (H_2-based) powered electric auto-vehicles (under development for near future introduction).

(c) Ships/Motors Boats /Submarines

— Diesel Engine Driven.

— Steam Engine (conventional/ reciprocating) driven, Steam turbine driven.

— Gas Turbine driven (hover-craft, high speed naval petrol boats, submarines).

(d) For Air crafts

— Aviation fuel operated piston engines, (but no more in vogue for larger Air crafts).

— Gas turbines (turbo prop, jet propulsion engines) operated with ATF(=Aviation Turbine Fuel).

(e) Rockets/Missiles

— Liquid(Liquefied O_2/ATF. Propane gas etc) and solid propellant fuel operated rocket engines on GT-operating system) (GT = Gas Turbine).

(5) Prime-Movers For Material Handling Equipment:

- For EOT Cranes/Jib cranes etc. — Electrical Motors;
- For Fork lifters — Battery operated Electric Motors, Diesel Engines;
- Inter Shop Transfer Trolleys — Electric Motors (Mains operated);
  — Battery operated Motors (for smaller size trolleys);

— Diesel Engine driven Winches;
— Steam Engine (reciprocating) driven Winches (**now NOT in vogue**);

- Conveyor System
  — Electrical Motors;
  — Magnetic Levitation (Electro-magnetic system based);

(6) Prime-Movers For Mining Operations:

- Electric Motors
- Diesel Engines
- Diesel Electric combination drives
- Steam Engines (at present, rarely used).
- Steam Turbines (blast furnace blower drive).

(7) Prime-Movers For OIL, Gas Exploration:

- As in(6) above; also small and medium capacity- (upto around 60Mws gas turbines in some locations employing large oil rig installations (both on-shore and off-shore).

NOTES: (i) Where sufficient quantum of electric power is not available, Diesel Engines and Steam Engines are common preferences.

(ii) At certain isolated locations, wind driven devices (wind-mills) and water-falls/high current water-flow driven mini/micro hydraulic turbines (also called Water wheel machines) are also employed for driving pumps, grinding of wheat for flour, besides driving small electrical generators.

(8) Some additional details about petrol engines and diesel engines as prime-movers:

Besides electric motors and turbines (steam, hydraulic, gas), the most commonly used prime-movers are petrol and diesel engines. In the following paragraphs, some more details about the application of petrol and diesel engines are given:

(A) Petrol Engines:

Petrol as a fuel being much costlier than diesel, the use of petrol engines are limited to lower bhp ranges for use as prime-movers in the following application: (**bhp = brake horse power**).

(A-1) Industrial Drives: (In the ranges from 0.6 bhp to say around 5 bhp)-normally for short-duration operations where electrical supply is not readily available:

* (i) For pumps, auxiliary drives, timber cutting/logging, small compressors.

*(ii) For electrical power generation:- mainly for portable small capacity generators in the range of 0.5 KWs to 3 KWs. (***NOTE: Such*** **generating sets are mostly for emergency power supply for short periods*).***

****NOTE: These petrol engines are mostly air cooled versions for operating for short periods.***

(A-2). Automotive Drives: (In the range of 0.6 bhp to around 63 bhp):- for driving:

(i) Two wheelers and three wheelers (Scooters, mobike, motor cycles, small 3-wheel Taxi Cabs etc), (0.6 bhp to 7.0 bhp); generally single cylinder air cooled engine.

(ii) Cars, vans, Jeeps etc. (7.5 bhp to 63 bhp)*:- generally 3 to 6 cylinder engines; (****NOTE: Bigger Luxury saloon and racing cars may have much higher bhp engines; their number is very limited as compared to the general purpose passenger vehicle in use anywhere);***

(iii) Petrol Engines for automobiles are also rated in volumetric capacity of their engine cylinders-such as 800cc, 1000cc, 1300cc (100cc-approx. correspond to 1 bhp).

(A-3) Outboard Engine Fitted On Small Motor Boats:

(1 to 3 bhp): Single cylinder engines. NOTE: Here again specially designed speed boats for racing (water-sport) purposes can have much higher capacity engines, and they can even deploy gas turbines/rocket engines to attain desired driving power and speed.

(A-4) General:

Petrol Engines in the lower bhp ranges are generally air cooled ones (say upto 3bhp). For higher bhp engines, engine cylinder block has either circulating water cooled jacket or uses a special grade of coolant liquid, depending on the design.

NOTE: In the last two decades or so, there have been certain revolutionary technological changes and advancements in automobile petrol engines to make them compact in construction, highly fuel efficient and also keeping the emissions within the internationally prescribed limits. The MPFI(=Multipoint Fuel Injection) system as well as electronic firing system have replaced the traditional carburation and electro-magnetic rotary spark-switching systems. Moreover, the 4-valve (per cylinder) and twin cam system have also introduced multi-stage sequential fuel-injection, and firing timing control enabling high level of fuel consumption to torque delivery ratio. These engines are comparatively silent with much lower vibration level.

(B) Diesel Engines:

Diesel fuel being comparatively cheaper, diesel engines find rather a wide range of applications. Diesel engines in different capacity-range are being used as prime-movers for:

(B-1) Industrial Drives (5 bhp to around 150 bhp): (other than power-generation)

- For pumps, rice husking mills, flour mills, line-shaft drives in small factories/work-shops (where electrical power is not available or facing power cuts), Jute/Cotton bailing mills, timber saw-mills, air compressors etc.
- While engines upto 15 bhp can be in single cylinder construction, engines in higher bhp ranges, depending on the speed and torque required, may come in 4 to 12 cylinder construction **(water cooled).**

(B-2) Automobile Drives (5.5 bhp to 160 bhp): (for cars, trucks, buses).

- Auto cabs 3 wheelers driven by single cylinder air cooled engines (from 5.5 to 12.0 bhp).
- Cars (10 bhp to 63 bhp, Trucks/Truck Trailers (60 bhp to 160 bhp), buses (60 bhp to 120 bhp), Vans/mini buses etc. (from 15 bhp to 60 bhp) **(Normal Ranges)**
- Cars, Vans, Trucks, buses, truck-Trailers may have 4 to 8 cylinder-engines, circulating water cooled, depending on size and bhp capacity.

(B-3) Mining and Earth Moving Vehicles/Equipment (50 bhp to 1000 bhp depending on size and load capacity required):

- Road Rollers, Bull-Dozers, Front-End loaders, hot-mix plants and road layers etc.
- Mining Shovels, Excavators, Drag-lines, Over-burden drills,
- Oil Rigs (On-shore, Off-shore).

(B-4) Mobile Cranes (25 bhp to 250 bhp) (can be used with torque multiplier)

(B-5) Locomotive Engines:

Diesel and Diesel-Electric: 150 bhp 6000 bhp depending on size and load to be hauled (***NOTE: Diesel shunters may have capacities from 150 bhp to 750 bhp; regular long-haul Diesel Engines can have capacities from 850 bhp to 6000 bhp).***

(B-6) Battle-field Vehicles:

(a) **Armoured Vehicles/Field Gun Carriers**- from 60 bhp to 250 bhp;

(b) **Battle tanks** (400 to 1000 bhp);

(B-7) Electrical Power Generation:

(i) For small capacity/portable Sets-from 5.0 bhp to 150 bhp (***NOTE: Small sets upto 15 bhp are generally in single cylinder construction – water cooled).***

(ii) Medium and Larger Sets-from 150 bhp to even upto 15000 bhp-;

(B-8) Marine applications:

For Motor launches/Motorboats, Ships, Naval Defence Vessels (Patrol Boats, Submarines, Frigates, tug-boats, destroyers etc.)- from 15 bhp to over 15000 bhp. ***(NOTE: Twin engines with twin screw propeller shaft arrangement, are deployed in some ships/naval vessels).***

NOTE: In all these cases, diesel engines of multiple cylinders (From 4 to 18/20 cylinders) (both, Vertical Line or V-Formation) are used with circulating water cooling system and forced ventilation fans.

■ ■ ■

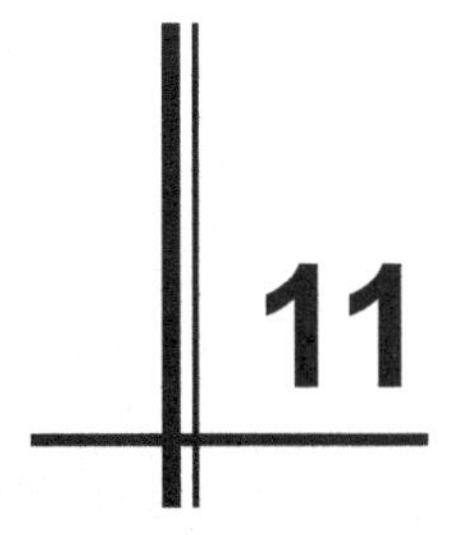

Engineering Materials, Engineering Drawings

"A diamond with a flaw is better than a common stone that is perfect".

—A Chinese Proverb

11.1 ENGINEERING MATERIALS

Engineering activities that lead to generation of products and services have to utilize certain kinds of materials: some naturally occurring, and some artificially created. Right from the concept to the stage of commissioning/ put to use, materials of different kinds, categories and sizes are required to be procured, processed and given required shapes for the desired industrial products and components.

(A) In the Conceptual and Planning Stage, engineers have to select, procure and use:

(i) Papers of different kinds for writing, preparation of drawing and documentation, modeling etc.; transparencies for reproduction, consumables for computers, other consumables and servicing materials.

(ii) Implements and gadgets for generation of engineering drawings, execution plans, and related reprographic and photographic processes like Ammonia printing, microfilming, sensidised photocopying, plain paper photocopying, photo-typesetting, word-processing (through PCs), computer graphics, computer imaging, CAD/CAM based documentation.

(B) For Production and Construction purposes, Engineering Materials used can be divided into two broad groups:

(i) **Metals and (ii) Non-Metals;**

While materials of both groups are being used for engineering industries, the use of metals are more predominant because of their unique properties like higher tensile and compressive strengths, ductility, high electrical and thermal conductivity, higher resistivity against distortion/deformation due to thermal and mechanical stresses, better resistance to environmental degradation..

Strengths and other physical and chemical properties of pure metals can be improved **by alloying with other elements**, and by the application of processes like heat-treatment, hot forging and cold working.

11.1.1 Metals and their Composition

(Metals which are commonly used in engineering application).

NOTE: ***For Mechanical Engineers and technicians, some reasonable level of knowledge on metallurgical properties of metals (metallurgical processes, metallurgical grading systems and standards, chemical compositions, and mechanical properties, selection criteria etc.) is imperative, particularly those working in the areas of engineering and design (structures, machines), quality control and testing, capital repairs, facility engineering, failure analysis, fracture mechanics.***

(A) **Metals:** Metals occur in nature as their chemical compounds and also get associated with other matters. As such production/extraction of pure metals is costly and difficult, **except for certain special applications** like high conductivity copper, Armco/Alnico iron for magnets and the like, **application of pure metals** is limited because of their higher cost. Therefore, metals and alloys of metals will contain impurities and their contents will be restricted to a maximum permissible limit, depending upon the usage. The impurities are always specified as maximum and the cost increases as the specified impurity level reduces.

For example: Sulphur (S) contents in:-

- Structural steels S = 0.04 Max. %
- High temp. bolt steels S = 0.02 Max.%
- Steam turbine rotor and blade steels; S = 0.01 Max.%

It should also be noted that an element which is an impurity in a particular metal can be an alloying element in another metal and under certain conditions can also be an alloying element in the same metal.

Example:- Sulphur (S) and Arsenic(AS) are impurities in steels. But 'AS' is an useful alloying element in brass, to counteract dezincification phenomena

(A Metallurgical phenomena viz; Arsenical brass condenser tubes:); 'S' **is added to steel for free-machining properties, but on the other hand, excessive 'S' contents beyond the maximum limits reduces the structural strength of steel and increases brittleness.**

(A-1) The **Alloying Elements,** in general, are specified in ranges but for production advantages, they are also some-times specified as minimum or maximum.

Wider the range, easier to manufacture, and hence cheaper will be the alloy. But with wider ranges, the properties and processing parameters will be wider and will lead to problems in processing. As a practical solution, a standard range is fixed for each element depending upon its effect on properties of that metal. Sometimes, still narrow ranges are specified for standardized processing in mass production or for critical components.

Example:- Alloy Steel grades(by carbon contents):

En 6: C = 0.6 max.%, (C=Carbon)

En 8: C = 0.35 – 0.45 %.

En 8C: C = 0.38 – 0.43%.

(A-2) Classification of Metals:

Metals for engineering applications can be classified as shown in the Chart below:

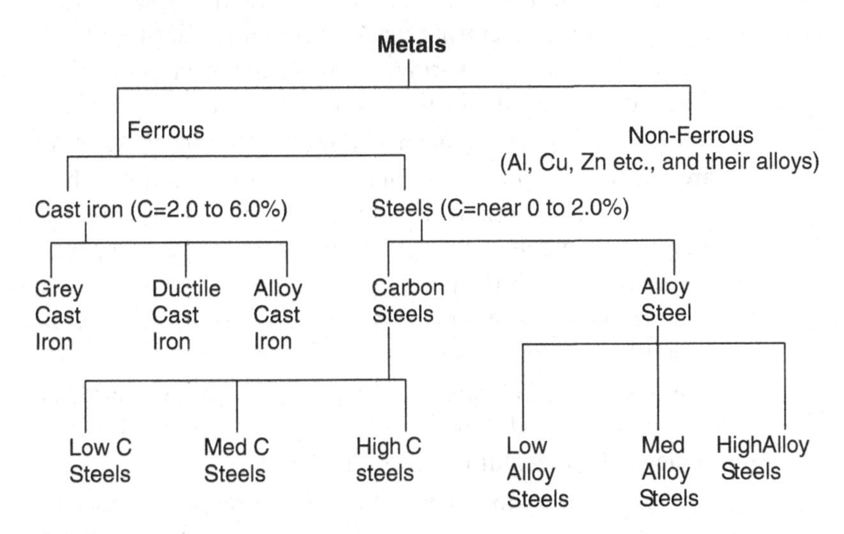

Non-ferrous metals and their alloys have limited use in engineering industries, when compared to ferrous alloys.

(A-3) Steels

In the Ferrous Group, the total production and consumption of low carbon steels and cast irons is very high when compared to other steels. This is mainly because of their cheapness and relatively simple processing techniques required. Even though the consumption of medium and high carbon steels and alloy steels is less, they are far more important than low carbon steels and cast irons, especially in engineering industries, because of their higher strength and special mechanical and chemical properties.

Ferrous Metals, particularly steels of different grades find wide spread use in industrial products specially in engineering industries making machinery, structural rolled and forged products etc.

(i) **Carbon Steels:** Depending upon the carbon content they are divided into Low (C<0.3%); Medium (C = 0.30 – 0.80%) and high (C>0.8%) carbon steels.

Low carbon steels are low strength steels. Their weldability is good and can be easily formed into different shapes. It is extensively used in beams, plates, channels, angles etc. Boiler steel also comes in this range, but with more stringent heat-treatment regimes to increase tensile strength.

Medium carbon steels can be heat-treated to achieve high strengths; welding and forming in this case become difficult as compared to Low carbon steel. This grade of steel is used in tubing, crankshafts, gears, heavy steel forgings, rails etc.

High Carbon steels are used only when properties like high fatigue-strength, cutting capability and wear-resistance are required such as in springs, cutting tools, engine-valves etc. Ductility and weldability are poor in high carbon steels.

(ii) **Alloy Steels:** Alloying elements like Ni, Cr, Mo, V, Al, B, W, etc., are added to improve hardenability i.e, to get uniform hardness even in bigger sections. In addition to improving hardenability, alloying elements give special properties like creep resisting strength at higher temperature, toughness, corrosion resistance, damping capacity, wear resistance, heat resistance, higher fatigue strength, high temperature hardness , etc.

All alloying elements, except Co, improves hardenability of steels but by different levels, whereas special properties are achieved by the addition of particular elements.

Example: Cr, Mo, V, W — imparts creep strength.
Mo — avoids temper brittleness.
Cr, Mo, W — gives high temp. hardness. (cutting tools).
Cr, Ni, Mo — Improves corrosion resistance.
Ni — increases toughness.

Alloy steels are always used in heat-treated condition to get optimum properties and to make use of improved hardenability. But majority of medium carbon steels and low carbon steels are used in rolled or normalized condition.

Creep Resistance Alloy Steels with proper grades and quantity of alloying elements (like Cr, Mo, V etc.) are used for construction of HP and IP casings and rotor forgings of steam turbines (ST), rotor forging and casings of gas turbines (GT), and blades of ST and GT (components which are subjected to temperature above 500^0 C). For other components of ST (LP casing LP rotor) exposed to lower temperature, low to medium carbon steels are used.

Depending upon the total alloy content, alloy steels are subdivided as **low alloy steels** (<5.0%); **medium alloy steels** (5.0 –10 %) and **high alloy steels** (> 10%). Majority of the alloy steels are of low alloy type and medium alloy steels are very few. **High alloy steels** are used for special purposes.

Example: **High alloy steels** — Stainless steels;
— High speed steels;
— High or High Ni heat-resistant steel;
— Heat Resistant steels, etc.

>>Stainless Steels Are of Three Types:

Austenitic (18 cr/ 8 Ni), **Ferritic** (low C, 16 cr) and **Martensitic** (higher carbon, 12 Cr). Austenitic varieties are not heat-treatable, Austenitic stainless steels are non-magnetic and got better corrosion resistance than the other two. (any steel having > 10% Cr is called stainless steel and their corrosion resistance is mainly because of Cr). **Martensitic and Austenitic** types have got good creep resisting strength, damping capacity and cavitation resistance. Hence they are used as hydro-turbine moving blades and pump impellers. Austenitic stainless steel with higher Cr and Ni contents are used as furnace parts because of their high temperature and electrical resistance.

(iii) **Grading of Carbon Steels and Alloy Steels with respect to carbon contents and alloying elements:**

Below is a list of some **SAE-AISI** designations for steel (XX in the last two digits indicate the carbon content in hundredths of a percent):

Grades of carbon steels and alloy steels	Alloying elements
• **Carbon Steels**	
10XX	Plain Carbon
11XX	Re-Sulfurized
12XX	Re-Sulfurized and Re-Phosphorized
• **Manganese Steels**	
13XX	Mn 1.75
• **Nickel Steels**	
23XX	Ni 3.5
25xx	Ni5.0
• **Nickel Chromium Steels**	
31XX	Ni 1.25 Cr 0.65 – 0.80
• **Chromium Molybdenum Steels**	
41XX	Cr 0.50-0.95 Mo 0.12-0.30
• **Nickel-Chrome-Molybdenum Steels**	
43XX	Ni 1.82 Cr 0.50-0.80 Mo 0.25

(iv) **Heat Treatment:** Heat Treatment is the heating and cooling operation that are carried out on casting and forging and other manufactured parts in order to achieve end property requirement. The heating temperature, cooling cycle are determined based on chemical composition of the component metal. In hardening operation of heat treatment process, steel parts are heated to a higher temperature, called austenitizing temperature. This temperature is above AC3 temperature, called the upper critical temperature, of Iron – Carbon diagram. In tempering operation, the hardened piece is heated to lower temperature, below lower critical temperature AC1, in order to achieve desired property.

(v) **Micro Structure:** Micro Structure plays an important role for the service performance of any metal component.

(vi) **Steel Classifications On End-Use-Basis:** The classification of steels discussed above is on the basis of chemical compositions. But steels are also classified on the basis of their use like creep steels, low temp. steels, carburizing steels, Nitriding steels, Boiler quality steels, Tool steels, Spring steels, free machining steels, forging quality steels, etc. Name of these steels indicate the use for which they are meant for. In the manufacture of these steels, special precautions are taken to guarantee their suitability for the specified end use. Even if the chemical composition and mechanical properties are same, steels of one type cannot be used for another type.

Example: IS 226, St 42 — General Structural steel.
BS 970, En 32 C — Carburizing steel.
IS 1875, CI IV — Forging quality steel.

DIN 17175, gr H II — Boiler quality steel.
ASTM A 516, gr LI — Low temperature steel.

(vii) **The Chemical Composition and Mechanical Properties** of all the above mentioned steels are almost same; but they are not interchangeable.

Depending upon the end-use, parameters like grain size, gas content, S and P content, internal defects, etc., are controlled during their manufacture. Because of this reason, materials should always be specified by their grade as well as the particular standard number; specifying only the grade may lead to selection of incorrect material. (S = Sulphur; P = Phosphorous)

(viii) **Cast Irons:** Cast irons are also mainly Fe-C (Ferro-carbon) alloy, but they are differentiated from steels by their higher carbon content and the presence of free graphite. Grey cast iron is the cheapest engineering alloy and can be machined very easily. It has got high damping capacity (used as machine tool beds) and good bearing properties, **but it lacks** ductility, weldability and formability, besides having low tensile strengths.

(ix) **Ductile Cast Irons:** Lack of ductility in grey cast irons is because of the embedded graphite flakes. So in malleable irons and spheroidal grey (SG) cast irons (ductile cast irons) the graphite shape is changed to nodules or spheres by special heat treatment and Mg-inoculation respectively. By spherodising the graphite, the notch effect of flakes is removed and distance between graphite particles are increased and hence the ductility of these cast irons improves. From the use point of view, SG – irons and malleable irons are very similar. Their cost is in between that of grey cast iron and steel. These ductile cast irons can be heat treated like steels although it is not mandatory. Ductile cast iron finds use in certain machinery parts where self-lubricating properties are necessary to minimize/avoid wear and tear.

(x) **Alloy Cast Irons**: They are special irons used for corrosion resistance, wear resistance, heat resistance, etc.; some of these grades can be non-magnetic as well depending an alloying element % age content.

Example: "Ni – hard" — Wear resistance. (Ni-hard rolls in steel rolling mills).
"Silal" — Corrosion resistance.

These find use in many applications in engineering goods/ products where both wear resistance and corrosion resistance properties are important

(A-4) Non-Ferrous Metals and Alloys

Al and Cu are the two main metals in this group. They are used in special applications. (Al = Aluminum; Cu = Copper)

(i) **Al:** Good electrical conductivity, low density and non-magnetic. Hence used as electrical conductors, aeroplane structures and in electrical instruments.

(ii) **Cu:** Good thermal and electrical conductivity, good corrosion and erosion resistance and non-magnetic.

Example: Arsenical Aluminum and Cupro-nickel tubes are used as heat exchangers tubes having good thermal conductivity and good corrosion/ erosion resistance.

High purity Cu (99.9%) is used as electrical conductors because of high electrical conductivity and can be easily made into wires and bars.

(B) **Non-Metals:** Many non-metals are used in engineering applications for different purposes. Basically non-metals are characterized by low thermal conductivity, high electrical insulation, non-crystalline structure, etc. **Polymer Chemistry** provided a range of synthetic materials called 'plastics', which are complex organic compounds. These pure plastics, often called as resins, have lower mechanical strengths. As such the strength of these plastics is often improved by suitable reinforcement. Some of the commonly used resins and reinforced plastics are briefly described below:

(B-1) **Resins And Plastics:** Resins are complex organic compounds, mostly derived from petroleum products and widely used for production of plastic, epoxies and composite materials. Polythene, polyvinyl chloride (PVC), acrylic plastics, polycarbonate, polysterene, polytetrafluoro-ethylene (Teflon), etc. are some of the commonly used plastics.

Alkyd resins, polyester, epoxy, phenolic and silicone resins are used with reinforcements.

Among these plastics, acrylic plastic (Perspex) is transparent and is often used in place of glass. Teflon is usually considered as a super-plastic due to its high temperature stability and electrical resistance to high-voltage applications. Polythene and PVC are generally used plastics for multipurpose applications.

(i) **Reinforced Plastics:** The basic resins like phenolics, polyester, epoxies and silicones are reinforced with paper, cotton fabric, glass fabric or cloth to improve the mechanical properties. Thus a range of reinforced products are developed. Among these, the following materials find extensive engineering applications:

(a) **Paper Reinforced Phenolic Resins:** The Materials obtained by hot pressing phenolic resin coated papers, usually known as phenolic paper laminates and are used in electrical insulation industry due to their excellent electrical insulation properties. By providing a decorative paper with melamine resin, a range of products called decorative laminates are developed.

(b) **Phenolic Cotton Fabric Laminates:** Cotton fabric coated with phenolic resin when pressed under heat produce phenolic cotton fabric laminates which are used in electrical industry as insulating boards, washers, tubes, spacers, wedges, supports etc. These are also being used now a days as civil construction materials (where there are low loads and stresses) like partitions, roofing of small sheds etc.

(c) **Fibre Glass Reinforced Laminates:** Laminates prepared with fibre glass reinforcement in epoxy, polyester, or silicone resins are used in engineering industry as these laminates posses very high mechanical strength and di-electric strength, and thus find use in electrical machinery, switch-gears, control panels etc.

(B-2) **Rubbers: Rubbers are organic polymers** having high resilience. Natural rubber is obtained from the latex of **rubber trees**. The hardness and strength of natural rubber can be improved by vulcanizing and reinforcement with fillers like cotton/nylon/acrylic fibres, graphite/carbon dust, silicon dust etc.

Apart from natural rubber, a group of synthetic rubbers are developed with better thermal stability and higher mechanical strength. These synthetic rubbers are used either independently or in combination with natural rubber. A few rubber grades are available for acid-resistant and oil-resistant uses.

(B-3) **Inorganic Insulating Materials:** Certain naturally occurring material viz,- asbestos, mica, quartz and silicon can be used as effective thermal insulation after due processing for shaping by bonding with synthetic resins like epoxy, acrylic compounds etc.

(B-4) **Synthetic Fibre:** Like glass wool/fibre acrylic laminates, teflon sheets etc are widely used as thermal insulation in furnace, boilers, steam pipelines, refrigerators, cooler etc.

(B-5) **Metallic Glass: The Material of the future:** This revolutionary material is made by combining metals like zirconium, titanium, nickel, copper and beryllium with glass and possess remarkable physical properties. Consider a strip of this shiny material as thin as paper. It flexes, and bends but does not tear how hard as you try, neither can you cut it with scissors or dent it. Drop a steel ball onto a slab of this material from a height of 2 feet and watch amazingly as the steel ball bounces back almost to the same height and goes on bouncing for more than a minute. On an ordinary steel or other surface, it would bounce only two or three times and come to rest. This is an entirely new type of material – called metallic glass – the material of the future. This new material can be used to build lighter and stronger things, from automobile bodies to aeroplanes, ships, bicycle frames, ortho-paedic implants, - the list is endless. Apart from its strength, metallic glasses have a distinct advantage over other types of materials. It can be melted and injection-moulded like a plastic to give it any shape and size in a mould. Similar metallic objects need to be made by machining, cutting, drilling, grinding-operations that are time consuming, costly and reduce the strength of the metal. Metallic glasses can also be readily made into a foam panel that is 95 per cent air but 100 times stronger than polystyrene. A sandwich made of 2mm of foam panel flanked by two thin layers of the metallic glass would be light, fireproof, bug-proof, rust-proof, bullet-proof, sound-proof and rigid and strong enough for car bodies, ships and aeroplanes.

(B-6) **Use of Timber and Composites in Engineering Industry**: Certain seasoned woods and other kind of woods/timber are used for making patterns for foundry, packing cases for shipment of products; certain composite materials are also used for making supports and fixtures where metals cannot be used.

(C) Cryogenic Processing pf Materials:

(i) **Cryogenic Process:** It is a process by which very low temperature conditions are created, say around 0°K (i.e. – 273°C). Under cryogenic conditions, metals and non-metals undergo substantial changes in their physical characteristics (Cryo-temperature range may vary depending on the type of materials); such characteristic changes in some metals are being usefully exploited by engineers and scientists for designing and fabricating special purpose equipment and devices for high-tech applications; for example, electrical superconductivity* (*near zero resistance conditions) can be achieved in some metallic alloys (Titanium-Neobium alloy) at around 4^0K (-269^0C) when enormous amount of electric current can be safely passed through the conductor without over-heating and damaging the same. The process is extensively used for the manufacture of moulds for ICs (Integrated Circuit elements) to ensure dimensional consistency by eliminating internal stress in the IC-Chip blocks.

(ii) **Cryogenic Technology:** Many important American defence and space establishments like NORTHROP AEROSPACE, NASA, RAYTHEON MISSILE, BOEING, COMET PRODUCTS,

HAMILTON STANDARD etc. are using a proprietary cryogenic process called “cryo tech” or “CPT” to effect dramatic improvements in the characteristics of metals and non-metals. This process advantageously affects every sphere of our engineering industry extending from Copper electrodes to compact disks. Infact, treated copper electrodes serve 400 percent longer; milling tools, drill bits, piston rings exhibit 300 to 2000 percent increase in life. Dies often outlast non-treated ones by 2x-8x (yet maintaining the dimensional stability and accuracy). Today, almost all razor blades are cryo treated increasing life from just a few shaves to as much as 25-30 shaves. Diamond cutting blades (granite cutting) last 200-400% longer. Ball and roller bearings show 200% to 700% improvement in load and wear characteristics. Nylon used in women’s garments last 4 times longer. Treated musical strings and compact disks exhibit deeper resonance.

Production in BAIRD engineering, USA, showed treated punches demonstrating a 357% increase in life, saving 72% in annual costs for the company. Broaches average 300% to 550% increase in life. Cryogenic treatment of Cobalt End Mills showed 26 times the wear life as against untreated ones. Deep cryogenic treatment of Gear Cutters helped a US company save 45 percent per part. Cryogenically treated carbide tools: performance increased by 400%. Treated Chain Saw Blades last 300% longer.

In CIBA-GEIGY, USA, cryogenically treated S-5 steel punches outlasted Diamond coated punches by 1200 percent. Copper electrodes Upped with Silver-Tungsten Alloy (proportion 1:1) used for switching 560 amperes at 440 V-Ac had a service life of 2 months;—after Cryogenic treatment, life has so far been extended to 20 months and the electrodes are still in service. Treated pure tungsten electrodes used in triggered spark gaps and treated Copper Tungsten electrodes exhibit a 1000% increase in service life due to a greatly reduced rate of electrode sputtering.

11.2 ENGINEERING DRAWING PRACTICES

(A) Engineering drawings are the language of the engineers and these translate the idea of the engineering into a real shape. A single sheet of drawing with relevant information (like title/name, dimensions, material of construction, scale etc.) indicated on it, not only replaces the need of several pages of descriptive texts, but also helps the reader to physically conceptualize and visualize the features of the subject of drawing such as machinery, layout of buildings, external and internal details of machines/structures/buildings, details of services like piping, electrical power supply system, ventilation and air-conditioning, control-circuitry etc.

For industrial application , the engineering drawings may be prepared mainly for the following purpose:

(i) Engineering/Design details, layouts, graphics etc.
(ii) Manufacturing process/construction-execution, quality control, assembly and testing, tools/Jigs/fixtures, packing/shipping etc.
(iii) For sub-contracted jobs (related to (ii) above).
(iv) For project execution/construction/planning etc.
(v) For procurement/purchase related processes.
(vi) Erection/installation, testing and commissioning (at the user end).
(vii) Trouble shooting, repair and maintenance.
(viii) Major Reconditioning and Retrofitting.
(ix) For customer/user appreciation/education, marketing.

There are standard procedures available on engineering drawing practices based on National, International and Company Standards, incorporating norms, rules and regulations.

Some important Indian Standards (issued by Bureau of Indian Standards) relating to engineering drawing practices are mentioned below. Engineers engaged in industrial and construction activities are supposed to be conversant with relevant ones:

- IS –10711 – 1983 — Sizes and layouts of drawing sheets;
- IS –11664 – 1986(RF – 1993) — Folding of drawings
- IS-6909-PartI-1985(RF-1990) — Lettering on drawings
- IS-10714-1983(RF-1993) — Methods of presentation (Projection etc.)
- IS-11065-Part-II-1984 — Isometric Projections;
- IS-11669-1986(RF-1993) — Dimensioning on drawings;
- IS-11775 — Modular Coordination, Symbols and Notations;
- IS-10713 — Scales on technical drawings;
- IS-11665 — Titles Block on tech. Drawings;
- IS-10164-1985(RF-1990) — Micro-copying of technical drawings;
- IS-696-1972 — Orthographic projection in 'First Angle' and 'Third Angle';

(B) Besides the standards issued by Bureau of Indian Standards, there will be organization specific standards to meet specific needs in the organization and those working there must get conversant with those relevant to their area of work and responsibility.

It is also very important for engineers to read/interpret and understand the drawings along with the symbols, notations and dimensions and be able to explain and guide technicians, workers and others concerned (including customers/users) to follow the underlying meanings and details for successful execution of jobs/projects/contracts.

In order to maintain consistency in the drawings prepared by several engineers at various places and to facilitate the component to get manufactured at any place, certain universally applicable rules and regulations are to be followed while preparing the drawings to avoid confusion in understanding while executing the drawings into real shape. These rules and regulations are laid down in the code of practices for general engineering drawing.

(C) **Sheet Sizes, Layout And Planning of Drawing Sheets And Folding of Prints (IS-10711-1983; IS-11664-1986 (Rf-1993)**

The preferred sizes of drawings sheets are given below:

(Figures in mm)

Sheet designation	Trimmed		Untrimmed	
	Width	Length	Width	Length
A0	841	1189	880	1230
A1	594	841	625	880
A2	420	594	450	625
A3	297	420	330	450
A4	210	297	240	330
A5	148	210	165	240

In arriving at the trimmed size of drawing sheets the following basic principles have been taken into consideration:

(a) Two successive preferred sizes of drawing sheets are obtained either by halving or doubling. Consequently the surface areas of two successive preferred sizes are in the ratio of 1:2.

(b) The formats of preferred sizes are geometrically similar to one another, the sides of each being in the ratio of 1:2. Surface area of the basic size A0 is one square metre.

(D) **Layout of Drawing Sheets:** The layout should facilitate the reading of drawings and made it possible for essential reference to be located easily when drawings are prepared by several offices. A standard arrangement should ensure that all necessary information is included and sufficient extra margin is left to facilitate easy filing and binding whenever necessary.

The title block is an important feature in drawing and should be placed at the bottom right hand corner of the sheet, where it is readily seen when the prints are folded in the prescribed manner. The size of the title block recommended is 185 mm x 65 mm. The size of the title block is uniform for all sizes. The following basic information shall be available from the title block.

(a) Name of the drawing (title);
(b) Drawing number and sheet number;
(c) Scale or scales used;
(d) Symbol denoting the method of Projection;
(e) Name of the firm;
(f) Dated initials of staff designing, drawing checking, standards and approving;
(g) The raw material specifications and the sizes of the material in case of component drawings;

It is preferable to show the folding marks on the drawing sheet so that they facilitate folding of prints in an easy and uniform manner.

(E) **Revisions:** It is essential that the drawings should record all alternations which are made from time to time. A convenient method of recording revisions is a table on the drawing giving the revision number or letter, date, zone, brief record and dated initials of approving authority. Care should be taken to see that the record of revision is tied up into the drawing in such a manner that it may be easily found.

When more than one sheet is required for a drawing of a part or assembly, and a particular sheet is from one such drawing, the numbering should show the number of sheets of the drawing as indicated below:

'Sheet 4 of 12' and so on, or, "Sheet-4/12" etc.

This entry should be made below the drawing number in the title block.

(F) **Additional Information On The Drawing Sheets:** Where appropriate the drawing sheet may include the following information either in the drawing or in the title block.

(a) Job or order number
(b) Treatment, finish, etc.
(c) Key to machining and other symbols and legends, scale
(d) Tolerance on dimensions not individually toleranced.
(e) Parts list (with item/part No., name/description, material specifications, quantity required, cross references etc).
(f) Other information including tool and gauge, jig and fixture references.

(G) **Scales and Scale Notations used on drawings:**

Full Scale, Reduced Scale, Enlarged Scale; for example:-

Full Scale	Reduced Scale	Enlarged Scale
1:1	1:2	10:1
	1:3	5:1
	1:5	2:1
	1:100 etc.	Etc.

(H) **Lines and Lettering:** Sizes, shapes, thickness, relative geometrical positions, readability, uniformity etc. are to be observed as per prescribed standards and practices.

(I) **Projections:** In an orthographic projections, a component can either be placed in **'First Angle' or 'Third Angle' disposition**. In India and most other countries (except those in North and South Americas who mostly follow third angle projection system) predominantly follow first angle projections system as a standard practice. However, IS-696-1972 provides for both systems. **International Standard Organization (ISO)** recommends first angle system of projection.

The system of projection followed must be clearly indicated on the drawing to avoid confusion.

(J) **Symbols, Notations:** Preferably as a single system conforming to National and Company Standards (properly interfaced) be adopted and legend on this be indicated on the drawing to the extent feasible (features like straightness, flatness, circularity, parallelism, perpendicularity, angularity, accuracy, tolerances, concentricity etc. be indicated in commonly understood symbols of standard practices).

(K) **Sections:** To show internal details, sectionalized view at the appropriate planes are drawn, generally in thinner line at 45^0 to the main out-line of the section. There are prescribed standard methods of indicating sectional views in various planes. For the engineering professional beginners, it is better to first get conversant with the system in vogue in the organization where they are working. This experience is useful in understanding National and International systems in vogue.

(L) **Dimensioning and Tolerances:** These must be indicated on the drawing as per standard practices. Dimensioning can be done in 'Aligned System', or in 'Uni-directional System', Dimensioning can also be done in 'Chain Dimensioning', 'Combined Dimensioning', 'Coordinate Dimensioning', Dimensioning Systems depending on 'Configuration of the component and practice in vogue.

NOTE; For more specific details, interest persons are expected to study books on Engineering Drawing practices and the relevant standards.

■ ■ ■

12

Power/Energy Supply for Industrial Units

"Engineers who do not take proper care for the resources placed at their command, also the costs and economics of their acts, are fanciful professionals that the company can ill afford to have them on it's pay-roll."

—Taken from a Lecture on Industrial Productivity

12.0.

Industrial units of all kinds require energy inputs for their smooth and sustained operations. This energy has to come from a cost-effective and economically sustainable regime to support viable operations.

Industrial Units require energy inputs mainly in two forms: Heat and Electricity.

12.1 HEAT ENERGY

For certain industrial processes, energy in the form of **heat** may have to be made available through suitable heat transfer media like hot water, steam, hot air, flue gases, etc. from the **heat generation sources** like furnaces, fire chamber of a boiler, or from heat-recovery devices providing heat as a by-product from certain other process plants.

While burning of fossil fuels like furnace oils, gas, coal, lignite, as well as biomass (agricultural wastes, Municipal garbage etc.) can be used in furnaces and fire chambers, **electrical energy/power** can also be used for getting heat energy through **resistance-heating system** provided sufficient quantum of electrical power supply at an economically sustainable cost is available on a reliable supply regime.

12.2 ELECTRICAL ENERGY

Besides heat energy (derived from burning fossil fuels and other combustible materials) required for process plants, automobiles, transportation and earthmoving machinery, **energy inputs in the form of Electrical Power** in substantial quantum are also required for various industrial drives and processes.

Now a days, **industrial drives** operate mostly on electrical power since it is the most prevalent, convenient, cost-effective and considerably environment friendly source of energy **(environment friendly at the user's end, but may not be so at the generation plant locations)**. Therefore, most industries prefer electrical power supply to meet their energy needs, excepting for the heat energy requirements for entire processes as mentioned above.

Please refer to Annexure-3 for Energy terms and definitions.

12.3.

***Fuel Supply Sourcing and Storage for Heat Energy as well as for Electrical Power Generation:** (*Fossil fuels).

(1) Coal/Lignite:

(a) **Sourcing:** These have to be sourced from the collieries through the marketing agencies of the Coal Mining and Supply companies; supply contracts are to be entered into for assured supplies on a regular basis.

(b) **Transportation:** Either through trucks or railway wagons. In case of Railway Wagons, a suitable railway siding with matching loading/unloading facilities may have to be installed inside or near the factory premises, or to be lifted by trucks from nearby railway station sidings.

(c) **Storage:** adequate and safe storage yards are to be created with matching handling facilities; effective steps to be taken against pilferage and fire hazards.

(d) **Quality Control:** there should be some arrangements and facilities to check the quality (calorific value per kg, ash and shale contents etc.) of the incoming supplies in conformity with the desired/promised specifications.

(e) **Distribution to consumption centres:** to avoid wastages and excessive consumptions (beyond specified limits), there should be some measurable parameters and matching devices to control; documented inventory control should be in place to facilitate quantitative control on receipts and issues.

(2) Oil, Gas, Naptha:

(a) **Sourcing:** They have to be sourced from the **Natural Oil and Gas Companies** through their marketing agencies; supply contracts are to be entered into for assured supplies on a regular basis, conforming to the desired specifications.

(b) **Transportation:** either through Vehicular Road Tankers, or through Railway Tank-wagons. In case of Railway tanker-wagons, a suitable railway siding with matching loading/unloading facilities are to be provided inside or near the factory premises, or to be lifted from the nearly railway station sidings by tanker trucks.

NOTE: In case of gases, transportation by vehicular road tankers should be preferred since such tankers can directly deliver the gas to the consumer's premises avoiding transhipment related problems at the railway station sidings.

(c) **Storage:** adequate and safe storage facilities with matching handling equipment have to be installed. Prescribed safety standards and regulations are to be followed for safe handling and storage and distribution; adequate precautions are to be taken for fire hazards and explosion, pilferage etc.

(d) **Quality Control:** as a consumer, it may be difficult to carryout any quality checks regarding grades, calorific value etc, and may have to depend on the good commercial ethics of the supplier agency, but should insist on specified quality by getting a quality certificate from them with each consignment. **However, large industrial units may have adequate laboratory test facility and expertise to carry out such quality checks.**

(e) **Distribution:**

- Distribution to consuming centres be controlled through a matching system and procedure(documented records for requisition and issue/returns etc.).
- **A system of regular checks** on wastages, pilferage, leakage etc. should be in place.
- Stock-control through inventory control system.
- Adequate precautions and measures be taken to avoid fire hazards and accidents at the dispensing points/centres.

12.4 ELECTRICAL POWER SUPPLY SYSTEM FOR INDUSTRIAL UNITS

For industrial units, electrical power can come from one or a combination of the following sources:

(1) **From the Grid System:** Transmission and Distribution system of the Electrical Power Generation and Distribution/Utility Services Company (ies):

* Either at HT(High Tension) or LT(Low Tension) or in a combination of both and a judicious mix, depending on the magnitude and type of loads as well as the distance between the factory/load centre and the grid system sub-station/power generation plant.

(2) **Captive Power Plants: (#)** Owned and operated by the company itself, mainly for assured supply/self-reliance vis-à-vis the rather un-assured/ erratic/insufficient power supply from the electrical supply/utility company. Captive power plants can be any or a combination of more than any one of the following generation methods:

(A) **Employing Diesel Generating Sets (DG Sets):** Diesel engines driving a Electrical Generator (now a days, predominantly an AC Synchronous machine) supported by various associated auxiliary systems.

(A-1) Although the cost of power generation per unit (Kwh) is normally higher than the purchased cost per unit from the grid system/ utility agencies, there are certain justifiable reasons for opting for installing DG sets to meet essential requirements of power for industrial units; **these are:**

(i) Power supply from the grid system as per requirement is NOT available, or is unreliable, (frequent interruptions, low voltage, low frequency conditions).

(ii) Access to grid system NOT available since the industrial unit is in an isolated place, much away from grid system.

(iii) Power consumption is rather low (A few KWs to may be around 1 to 3 MWs) and intermittent/short period demand; also the cost can be absorbed in the overall viability of operations in a sustainable basis.

(iv) For meeting only the peak-load demand when the grid system fails to supply required quantum of power, and put a quantum restriction by the power company..

(v) For meeting emergency power demand for some critical operations when the grid fails or grid power is cut / interrupted.

(A-2) DG sets are available in the market in various sizes, from single cylinder 5 KWs capacity (single cylinder DG sets normally up to 15 KWs capacity) to around 6000 KWs in multiple cylinders (up to 16 to 18 cylinders). Now a days, silent DG sets with much lower noise and exhaust pollution levels are available in the unit capacity upto 2000 kws, with battery bank operated cranking motor start. There are DG sets of unit capacity up to around 200 KWs with auto-start and auto-loading facilities (auto start to full loading in 15 to 20 seconds).

(B) **Captive Steam Turbine Generator Sets (STG Sets):*** When the electrical load is substantial, DG sets may not be techno-economically /commercially feasible for large scale power generation for a continuous/prolonged periods. In such a situation, STG-sets present a viable option, but again subject to conditions relating to feasibility and practicability with respect to (**mainly**):

- Assured fuel supply sources (furnace oils, coal, lignite, natural gas, naphtha etc.) at cost-effective regimes.
- Availability of adequate land/space and water supply sources.
- Sorting out pollution control/environmental issues.
- Transportation , storage and handling of fuel.
- **And most importantly,** cost of such a STG plant: magnitude of Capital Investments and operating costs for smooth operation for assured power generation and supply.

****NOTES:* (i) *The company has to carefully examine and analyze various related techno-commercial/techno-economic issues for getting power from the grid system vis-a-vis going in for the option of installing DG sets.***

(ii) STG Plant is a complex set up with a number of auxiliary systems; this will not only need heavy capital investment, but also supporting organization for operation and maintenance, besides substantial level of operating costs.

(iii) It should not happen that STG plant operation DIVERTS substantial degree of management attention at the cost of the core-business/production activities, getting lesser attention than essentially required.

(C) **Co-generation Power Plants**: This is **basically a STG plant** utilizing waste-heat/by-product heat, or surplus steam from any existing process plant in the Industrial Complex, with fired or unfired **Heat-Recovery Steam Generator (HRSG).** Steam thus made available (normally low parameters steam) for operating back-pressure (non condensing type) steam turbine to run an **electrical generator** to generate power. Such plants are rather small capacity ones in the range of 500 KWs to 5000KWs, depending on the quantity (tonnes/hr) and quality (temperature, pressure) of steam available. These power plants are much less complex in construction and layout features and simpler to operate, but cannot generate power on a large scale.

NOTE: Possibility of such cogeneration exists only when the industrial unit has sufficient quantity of waste heat/surplus by-product heat available from the existing process plants.

(D) Other Types of Captive Power Plants:*

(D-1) **Gas-turbine Generator Plants (both open cycle and combined cycle)**: This is again a complex set-up and entails heavy capital investment and operating costs, besides other resources constraints; therefore NOT normally recommended for industrial units, particularly small and medium size enterprises
***(Not An Economically Viable Option).**

(D-2) **Solar and Wind Power:*** These **Non-Conventional** power generation sources(**Solar Photo-Voltaic Panels**), because of many limitations associated, are not very useful for **INDIVIDUAL industrial units** unless such systems are established near the load centre by a Utility Company and the power is made available in a cost-effective and dependable manner; **at one location** more than a few MWs of power may not be available.

However, in the wind power sector, an increasing number of such sets are being installed in different parts of the country, in the range of a few KWs to around 2500 KWs. In India, the largest wind turbine generator so far installed is around 2500 KWs. However, the largest unit capacity of wind turbine set so far installed anywhere in the world is around 8000 KWs (IN Europe);

***Not suitable for individual industrial units requiring power in substantial quantum.**

(D-3) **Biomass Power Plant:*** There are normally two methods of utilizing Bio-Mass for power generation:

(i) Bio-mass Digester Plants generating biogas which can be used as domestic cooking fuels and for localised lighting through the use of gas stoves and gas lighters respectively.

(ii) Bio-mass generated from agricultural wastes and **municipal garbage** can be burnt in a fire box of a boiler (normally flue gas type) to generate steam to run small steam turbine generator(STG)* sets (normally not more than 5 to 15 MWs in one location). **Proper emission control devices will have to be incorporated for complying with pollution control regulations.**

NOTE: Such STG plants may not be preferred as captive power plants by the industrial enterprise since availability and continuous supply of biomass fuel materials will be difficult to manage. They are best suited for smaller urban areas to be managed and run by utility services agencies/municipal agencies for local load centres with limited demands.

(D-4) **Captive Hydro power plants: This is a rare occurrence;** a few industrial enterprises have captive hydro power plants installed in the vicinity area taking advantage of the existence of a suitable water falls/ stream, but such power plants are generally of small capacity ranges from 500 KWs to around 3000 KWs. Capital expenditure can be quite high.

■ ■ ■

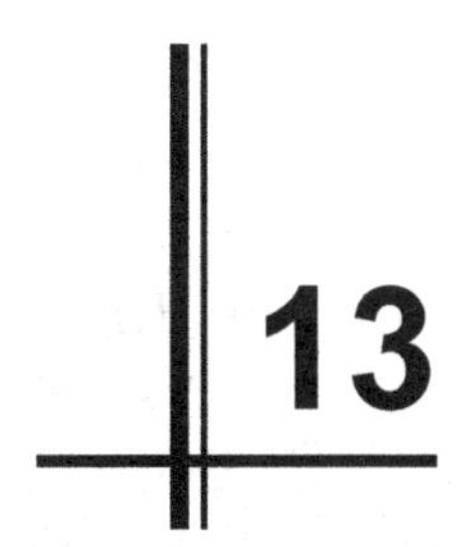

Selected Miscellaneous Topics

"Without knowledge, the world is bereft of culture.
And so we must be educators and students, both".

—Roberta Bondar, Astronaut

A reasonable level of familiarization with the below mentioned topics may help the Young Engineers to prepare themselves towards acquisition of competence for the tasks of **Industrial Administration,** a way forward to career growths; we are presenting below elaboration on the same.

13.1 SYSTEMS AND METHODS INSTRUCTIONS AND PROCEDURES

Industrial Units, big or small, have to adopt and follow certain systems, methods and procedures in their day to day working for smooth and efficient operation.

A System may be defined as a set of instructions, combined in some fashion, directed to some purpose. System explains how to perform specific activities or tasks: who is to do what, when and the relevant organizational mandate to be followed for the same.

Procedures are further elaboration of Systems from the point of view of actual implementation process to be followed, clearly indicating the specific roles and responsibilities for various agencies/individuals to achieve the end results desired.

(1) Systems are formulated and brought into practice in an organization to:

- Provide Role-clarity.
- Serve as a Common Language for working.

- Avoid confusions and conflicts.
- Improve Work-culture.
- Support Efforts for achieving **overall Performance Excellence.**
- Avoid adhoc-ism and personal whims.
- Helps increasing awareness for adoption of better and healthy working systems in the organization across the board.

(2) **Systems and Methods** can be termed under different titles and terminologies in various organizations depending on the organizational culture and preferences, but the purpose basically is the same:

- **SMI (Systems and Methods Instructions);**
- **WMI (Works and Methods Instructions);**
- **OMI (Organization and Methods Instructions);**
- **SP (Standard Procedures);**

---------- the like of these.

(3) **Systems and Methods as relevant to Industrial and Business Organizations, can be broadly classified as under, depending on functional applications:**

NOTE: The list below is representative and not exhaustive as there can be large number of functions in a large complex organization;

- **General Organizational Systems** : Relating to General Administration, Establishment matters, Personnel Policies (HRM, HRD) and Industrial Relations, systems related to TQM, EMS and activities related to, security, Public Relation.
- **Engineering Systems** : Relating to Product Design and Engineering functions and associated work instructions and authorization.
- **R&D System** : R&D objectives, R&D strategies, procedures for selection, initiation, authorization, investment and budgeting, procedure for closing/abandoning, record generation and data storage, commercialization etc.
- **Production Systems** : Relating to activities on production/ manufacturing/assembly/testing/packing/ dispatch .

- **Plant Maintenance and Upkeep System** : Relating all activities and processes maintenance/repairs, upkeep, replacements, retrofitting of plant and machinery, buildings, other infrastructural facilities.
- **Quality Systems** : Relating to Quality Assurance, Quality Control Checks, ISO-9000-QMS and TQM related activities.
- **Financial Systems** : Relating to Financial Management and Accounting, costing, PSL, Pay Roll, sales costing, sales billings, cash management, audits, dealing with Banks/FIs; budgeting and budgetary controls, preparation of **balance-sheets**.
- **Commercial Systems** : Relating to commercial/sales/pricing processes, customer services, invoicing, billing and sales proceeds realization, dispatch-related matters, customer project monitoring and follow up, tendering, order booking etc.
- **Marketing Systems** : Marketing objectives and strategy, procedures for various segments of market, procedure for dealership accreditation, marketing outlets and distribution, procedures for tie-up with other agencies, leasing procedures.
- **Trading System** : Trading objectives and strategies, procedure for commodity procurement and distribution, dealing with dealers, procedure for customer dealing, sales proceeds realization, invoicing etc.
- **Exports Related Systems** : Export objectives and strategies procedures and paper work system, in conformity with National and International trading procedures, incoterms etc.
- **Materials Management Systems** : Relating to Materials Management aspects like procurement/purchase, receipts, storage, conservation, distribution, disposal, vendor evaluation and registration, vendor development, sub-contracting/off-loading, **inventory control** etc.

- **Computerization Systems** : This relates to all areas of operation as listed above, and formulated for enabling computerized-systems of working (inputs, output, corrective, record keeping etc).
- **Collaboration/Joint Venture Related Systems** : Objectives and strategies, procedures for transactions, role definition and division of responsibilities, profit/loss/ liabilities sharing, equity and technologies sharing, (as the case may be), market sharing, buyback terms, royalty payment procedures etc.
- **Facilities, Infrastructure and Assets.** : Systems and procedures relating **Asset Management** planning, acquisition/ procurement, installation, upkeep, replacement, disposal, re-conditioning, capital investment, record keeping/asset registers etc. for all types of capital facilities/Fixed Assets, as well as other small assets. (accessories, tools, tackles, fixtures, jigs, dies etc.).

(4) **Technical Services Functions: A discussion on application of Standards, Systems and Procedures will be rather incomplete without appreciating the role of Technical Services Function** which, in most enterprises, act as the nodal agency for Standardization and other Technical Services.

Technical Services Function in an industrial/manufacturing/ trading organization may generally deal with the following activities:

(i) Provide reprographic facilities for documentation/drawings (including Micro- filming) etc.

(ii) Provide Archival Services for documentation/standards/ reference materials.

(iii) Provide Library Services for technical and non-technical books, magazines, periodicals etc.

(iv) Keep custody for various National and International Standards and make copies/extracts available to users as and when needed.

(i) Arrange to procure updated/latest issued **Standards** for the benefit of users.

(ii) Coordination for the formulation and circulation of various corporate/plant standards amongst the units/departments/ branches.

(iii) Keep liaison with the National Standards Organization for obtaining information to keep the **standards directory updated**.

Also coordinate for giving comments/suggestions to the National Standards body as and when new draft standards are formulated and circulated.

(iv) Adopt computerized working to the extent practicable for quick dissemination and retrieval of information on standards and related matters, and also for archival inventory records/status etc.

NOTE: ***In some organizations, Technical services function may also deal with activities like Corporate Affairs, Contracts Coordination, legal cell, liaising with external agencies/collaborators/joint venture partners, engineering and commercial coordination with major customers/consultants for major projects, departmental training coordination etc.***

13.2 PRACTICING EXCELLENCE IN FACTORY MANAGEMENT: ADOPTING THE 5S WORKING CULTURE

The **5S** working concepts were developed in **Japan** to achieve orderliness and excellence on the shop-floor in the factories to make them world-class establishments.

The **5S** factory management techniques aims at instilling in the workers at the shop-floors, a sense of disciplined work-culture which rests on the logic that cleanliness, orderliness and discipline at work sites are basic requirements for manufacturing high quality products, with little or no waste and with high level of productivity.

The 5S stands for five Japaese Words: Seri (clearing up), Seiton (organizing), Seiso (cleaning), Seiketsu (standardizing), and Shitsuke (training and discipline).

5S System of work-culture operates broadly in three domains as under:

(1) **Active 5S:** The propagators of the techniques recommend that before putting the **5S** to work, it is advisable to take photographs (colour) of the worksites as they are existing and keep them in record, so that these can be used later for comparison purposes with those photographs taken after the **5S** working system has been put to full practice. This will help adopting colour coding for creating a positive impression on the concerned work-force.

Let us now understand the significance of the five principles of ACTIVE 5S Working concepts:

(a) **SERI: Clearing up the area:** This process aims at identifying at the work site, items which are needed there and the items which are unwanted/ superfluous. This is to be done after clearly defining and deciding the standards for 'what is really needed' and 'what is not needed' to avoid difference of opinion and excuses. The un-wanted/not needed items to be marked with red rags, so that anybody can see what it to be eliminated from the work-site. For a bias-free identification process, an independent agency/person (not responsible for the concerned work-site but adequately conversant with the working requirements of the specific work-site) be assigned the task.

(b) **SEITION: Organizing: Organizing is the most far reaching one amongst the 5S steps.** After clearing up the area of unwanted/ superfluous items, the next step is to organise the area for each and every machine/work-sites by putting the materials, tools, tackles, drawings /documents etc in dedicated storage area/ storage cabinets/ bins etc and earmarking and leveling by large visible letters, numbers, sign-boards for the specific machines and work-sites. Such identification levels should be in language and colour codes which can be easily understood by all concerned. The system further recommends that the storage bins, cabinets etc. be open type without doors.

The best approach for the shop management/supervisors is to switch from the **'remedial organizing'** to **'preventive organizing'** by raising questions like why this mess-up happens and how something can be done about it.

(c) **SEISO: Cleaning:** A regular cleaning system has to be put in practice keeping the three broad categories of cleaning targets : storage areas, machines and equipment, and the surroundings. The cleaning efforts have to be organized in **the following recommended manner to consolidate the procedures:**

(i) Divide the work-sites into small areas and identify and list out all the cleaning work that are needed to be done.

(ii) Draw up a cleaning responsibility chart and prepare a cleaning schedule and display the same prominently at the respective work-places so that everyone can see.

(iii) Adopt a shift-system for the areas of common interest/ joint responsibility of various working groups.

(iv) Propagate the habit of daily five-minute cleaning schedule at all places; although the time is too short, but over a period of time, things will improve and efficiently done cleaning work will show more than expected results.

(d) **SEIKETSU: Standardization:** Standardize a spotless work place and impress upon all concerned to put in a little effort to keep the work-place clean and tidy all the time. The standardization effort must **keep three 'NO-Principles' in view** for successful implementation : **No un-necessary items, No mess, No dirt**;

Re-checking and appraisal are vital to give the system an impetus for perpetuity.

(e) **SHITSUKE: Training and discipline:** Training of the work-force which should also inculcate a sense of discipline in them, is an essential part of the practice of **5S** work-culture. This training will be an inter-active process which will not only involve preaching but will also **recognize the worker's participation** in constructive criticism and decision making so that the system gets wider acceptability. The idea is to create a work-place where problem points can be easily recognized and remedial action taken.

(f) **After the 5S working concepts have been successfully implemented**, it is now time to take a set of colour photographs of the present situations and display both the sets of colour photographs of **'Before 5S'** and **'After 5S'** positions to demonstrate to all concerned the amazing difference that has come about which resulted in transforming the factory towards a world class establishment.

(2) Effective 5S: In order to sustain 5S work-culture, the following aspects are to be taken care of:

(i) **Implement Visual Control:** by way of photographic display as stated in para–13.2.1 above and also circulate the results through handouts for wide publicity.

(ii) **Introduce a Reward-Scheme** for best performing work-places.

(iii) **Avoid, by all means,** becoming complacent and slackening up.

(iv) **Constantly strive** to make **5S** work-culture habitual.

(v) **Control stock levels:** by gradually working to decrease inventories without disrupting production and establishing an optimum level.

(vi) **Make it easy** to return things which are surplus or not needed at the work-place: simplify paper work-procedures.

(vii) **Put names and numbers** on all jigs and tools and out line their shapes at their precise storage positions. **Ideally,** tools should be stored besides the machines with which they are to be used and according to the sequence of work operation.

(viii) **Organize everything by a suitable colour code** for easy recognition at a glance.

(ix) **Always clean and check.**

(x) **Maintain standards** throughout the company; discipline has to begin with following strict standards.

(xi) **On the spot correction** should be done whenever there is some sign of disorder, and this should be got done through the area group-leader for effective results, since his commitments will be crucial for effective **5S** work culture.

(3) Preventive 5S: Once the **5S** work-culture has taken root and becomes habitual, the factory management has to assiduously work for maintaining the standard achieved. This will require certain preventive steps to be taken with determination and with a sense of sustained emphasis. **The preventive 5S will lay stress on the following:**

(i) Avoid un-necessary items and prevent their coming back by rationalising systems regulating supply and consumption.

(ii) Shift from lot-production to leveled-production and from lot-deliveries to multiple small-lot deliveries; this will reduce chances of excess material piling up at the work-place.

(iii) Avoid disorganization in storage; **three basic points to be remembered about storage : Where, What and How Many.**

(iv) Create a system which allows only the necessary items to be stored at the work-places.

(v) Ensure cleaning regularly and do not tolerate accumulation of dirt and scraps, wastes, residues etc. at the work-places.

13.3 JISHU HOZEN' (AUTONOMOUS MAINTENANCE): A TACTICAL CONCEPT OF PLANT MAINTENANCE

During the years after the World War–II came to an end (in 1945), Japanese government and industrial enterprises, like those in the European Continent, embarked on a massive urban and industrial reconstruction effort. In the decades that followed, Japanese industries and business enterprises not only achieved rapid progress and growths in respective area, but also, in the process, evolved and developed many new revolutionary management concepts and philosophies which provided added impetus to their success and enabled the country to attain the status as an industrialized high growth enterprising nation. As the success stories spread across the world, Japanese concepts of industrial and business management got propagated and to a large extent got accepted/introduced in the enterprises in other industrialized countries in the Europe, Asia, and even in the Americas (USA, Canada, Brazil etc.). The process is still on and a cross-fertilization of Japanese management concepts and styles and those in other industrialized nations is taking place on a larger scale for mutual benefits of experience sharing and collaborative relationships/arrangements for enhanced economic, trade and technological developments.

Besides many new concepts developed and propagated for activities in the area of production/manufacturing, productivity, quality management such as 'Quality Circle', 'Zero-defect', '5-S', 'TQM (Japanese version)', 'Group-Productivity', 'Kaizen' etc., Japanese industries also developed many new work culture concepts in the areas of plant maintenance activities to improve availability, reliability and hence productivity as well. **'JISHU-HOZEN'** or 'Autonomous Maintenance' is one of such concepts developed for a 'Tactical Plant Maintenance' strategy which became quite popular and acceptable in the areas of factory management since it aimed at ensuring better upkeep and maintenance of plant, machinery/equipment which are so vital and critical for production and productivity, besides assuring quality.

>> **JISHU-HOZEN** activities focuses on the equipment to make it perfect i.e. Zero Break-down and Zero-defective products processed by the machine tools/equipment on the shop floors.

It builds a relationship between the man and machine **through a sequence of 7 steps** (refer (2) below) whereby the machine operator owns the machine to the extent of taking care of its routine maintenance activities as well, besides his assigned manufacturing jobs.

(1) **What is Jishu-Hozen?** JISHU-HOZEN, A Japanese word for Autonomous Maintenance, implies that the operators themselves maintain their machines in highest standards. These operators are trained to do so through the structured approach of **7 steps of Jishu Hozen** as enumerated below. Each steps has a focus, an activity to be carried out and certain results to be achieved :

(2) **Autonomous Maintenance Methodology—the 7-Step Sequence:**

(i) Perform initial cleaning and inspection;

(ii) Eliminate contamination sources and address inaccessible places;

(iii) Develop tentative standard for cleaning, inspection and lubrication;

(iv) Conduct general equipment inspection training and develop inspection procedures;

(v) Conduct general process inspection autonomously (i.e. self-inspired);

(vi) Set standards and manage the wok place;

(vii) Pay attention to the ongoing autonomous maintenance and advanced improvement activities;

Mainly operators and cross-functional teams with support and training provided by maintenance personnel, production managers and engineering staff, carry out these activities.

(3) **Goal of Autonomous Maintenance:** To develop operators to maintain the equipment and wok places and follow uniformity in implementing shop-floor activities to eliminate forced deterioration

on equipment. Since operators have to ensure productivity and quality, they are bound to take better care of their machines to maintain their credibility and ensure earning of incentives often given in monetary terms.

(4) **Why Jishu Hozen is significant?** "Jishu Hozen" means activities of the machine operator are also focused on the upkeep of the machine he works on, by personally conducting maintenance activities, including cleaning, oiling, retightening and inspection. This in turn raises production efficiency to its limit. Such activities will prevent forced deterioration of the machine tools and other production equipment.

Activities of Jishu Hozen aim to eliminate:

(viii) Machine Failures,
(ix) Minor Stoppages,
(x) Defects in the machine and its output,
(xi) Other losses (e.g. energy losses, lube oil wastages, rejection, man-hour losses etc.).

*Jishu Hozen activities restores the equipment to their desirable forms, maintain them, and improve them, and at the same time develop personnel who are skillful at operations of the equipment as well as its improvement and upkeep.

*Jishu Hozen is the operator's own work itself. It is carried out under the shared approach through small-group activities closely unified with the factory management structure.

*Jishu Hozen itself signifies the constitutional improvement in the relationship between the personnel and the facilities. This entails the changing of worker's ways of thinking and behaviours by replacing the traditional concept of division of labour between operation and maintenance. This concept aims at replacing the thinking towards the work culture of "owning and protecting one's equipment by oneself."

(5) **Concluding Remarks:**

(i) To translate this philosophy into action, the development of personnel who are skilful at machine tool/equipment operations are persuaded further to get conversant with the structure and function of the facilities, as well as to acquire maintenance skills to effect improvements in upkeep and functioning, **thus bringing in attitudinal reform in the concerned personnel.**

(ii) It is also very important to bring about structural and operational improvements of the machine tool/equipment currently being used. In "Jishu Hozen", as the first step to "protect one's equipment by oneself", work should be started with improvement of basic conditions of the equipment being used;

Cleaning, Oiling, and (re)Tightening; i.e. (COT). Then, based on the step method, structural improvement of machine tool/ equipment should be effected so that the equipment will be in the form as it should be.

(iii) As a result, **Overall Equipment Efficiency (OEE)** will improve greatly, and the machine operator personally experiences the effect of "Jishu-Hozen", so that his attitude will be changed so as to protect his own machine tool/equipment by himself. **Thus, in "Jishu-Hozen", the constitutional improvement of personnel and equipment proceed simultaneously**.

(iv) "Jishu-Hozen" is a self management activity by small work-site groups. Keys to the activation of small groups are said to be **"morale, skills, and places for actions**."

(v) Through the development of "Jishu-Hozen" work culture, the machine operator's willingness and skills increase and encourage him to "protect his own equipment by himself." Literally, he will be able to manage his work autonomously. In "Jishu-Hozen", a small group set its goal for the attainment of corporate objective by itself; it uses machine tool/equipment as its own tool for the attainment of the goal; and it tackles production and improvement activities. Therefore, unprecedented, remarkable results can be achieved and activation based satisfaction of desire for growth and self realization will result. "Jishu-Hozen" is the practice of participation in factory management and respect for human endeavours.

13.4 LEGAL FRAMEWORK SYSTEM FOR BUSINESS AND INDUSTRY: A BRIEF ON LEGAL REFERENCES/ RELEVANT LAWS/ACTS: COMMONLY APPLICABLE AND KNOWN TO TRADE, COMMERCE, BUSINESS AND INDUSTRY (An Awareness Brief)

Engineers and Business Management professionals working in business and industrial establishments should also know about the existence of the following, so that in case of exigency, can take some remedial/preventive preliminary action before approaching Legal consultants/Experts on the subject for further follow-up action:-

NOTES: (i) The Laws/Acts go on being amended, as well as new laws are also introduced from time to time; consult legal advisors to find out the latest applicable.

(ii) ***The idea of presenting this topic is NOT to make the professionals as legal experts, but to make them aware of such legal requirements.***

A. SPECIFIC LABOUR LAWS:

- The Apprentices Act 1961.
- The Contract Labour (Regulation and Abolition) Act. 1970
- Misconduct/Dismissal/Discharge and Domestic Enquiry Procedures
- Minimum Wages Act 1948
- Workmen's Compensation Act, 1923
- The Employees' Provident Funds and Misc. provisions Act, 1952
- The Employees' State Insurance Act, 1948
- Equal Remuneration Act, 1976
- The Industrial Employment (Standing Orders) Act, 1946
- The Maternity Benefit Act, 1961
- The Payment of Gratuity Act, 1972
- The Payment of Wages Act, 1936
- The payment of Bonus Act, 1965
- Factories Act, 1948
- Industrial Disputes Act, 1947
- Shops and Establishments Act (Each State may have separate Acts)
- Trade Union Act, 1926

B. TRADE UNIONS:

There are a large number of Trade Unions operating in various Industrial Establishments, Agricultural Industries, Coal and Mining Sectors, Railways, Insurance and Banking Sectors etc, most of them directly or indirectly get affiliated to the following well known **major National Trade Union Organizations:-**

- **INTUC = Indian National Trade Union Congress**
- **AITUC = All India Trade Union Congress**
- **CITU = Centre for Indian Trade Unions**
- **BMS = Bhartiya Majdoor Sangh**
- **HMS = Hind Majdoor Sabha**

>> Some Other Related Laws Covering Business, Trade and Industry (A selected listing)

- Companies Act 1956/Companies (Amendment) Act 1974 and 2013;
- Indian Explosives Act 1884 and subsequent Amendments.

- Indian Boiler Act 1923 and subsequent Amendments.
- Environment Protection Act 1986.
- Air (Prevention and control of Pollution) Act, 1981
- Water (Prevention and control of Pollution) Act, 1974

>> **LAW OF CONTRACTS:**

"The Indian Contract Act, 1872" covering aspects like agreements, contracts, proposals, discharge of contract, breach of contract, guarantees, indemnity, pledge etc, and related legalities.

- Commercial Laws, Laws of Partnership, Law of Arbitration, Laws of Negotiable Instruments..
- Sale of Goods and hire Purchases.
- Law of Insolvency.
- Indian Electricity Act, 1910-and as Amended by Act 50 of 1991.
- Indian Electricity Rules 1956 (updated till 1992).
- Right to Information Act,2005.
- Indian Copy-Right Act of 1914, of 1957, and Amendments on the same Introduced in 1983, 1984,and 1999.
- IT Act 2000 (Cyber Law);
- Income Tax Act,1961.
- Essential Commodities Act,1955.
- Insolvency Act,1920.
- Societies Registration Act,1860.
- The Customs Act,1961.
- The Insurance Act, 1963 and 1972.
- Partnership Act, 1932.

■ ■ ■

14

Application of Standards in Business and Industry

"Shoot for the Moon, and if you miss, you will still be amongst the stars"

—Les Brown

14.1 APPLICATION OF STANDARDS

Application of Standards is very important , not only for engineering and design procedures, but also for various other activities and processes that are to take place in construction projects as well as on the shop-floors during manufacturing: assembly and testing, stage-wise inspection and quality control, packing and dispatches to customer, besides ensuring acceptable quality of input materials, material testing, documentation, record keeping, reference-coding etc.

The **Application of Standards** may also assume significant role in engineering goods trading houses, consultancy firms, construction companies, inspection, and contracting/sub-contracting agencies.

For **Universal Acceptability** in the market, it is essential to follow certain common standards which may be National like BIS/IS in India and International ones like DIN,BS, ASME, ASTM, IEC, API, ISO etc.

Even in the individual organization there can be Corporate Standards and Plant Standards for streamlined working within the organization.

For the practicing engineering and business administration professionals, a reasonable level of knowledge of applicable standards / codes and their usage/ relevance are essential, as without such knowledge, practice of engineering and techno-commercial matters will definitely be less than imperative necessity; lack of such knowledge will be an impediment and troublesome for business dealings.

14.2. Standardization and Regulating Bodies

For the benefit of the young engineers, the following is a representative listing of applicable standard-series and standardization/regulating bodies, widely used/ known the world over:

ISO	=	International Standards Organization, (Geneva)
BS	=	British Standards (UK)
DIN	=	Deutsche Industrien Normein (Germany)
ANSI	=	American National Standards Institute (USA)
BIS	=	Bureau of Indian Standards (India)
ASME(USA)	=	American Society of Mechanical Engineers (USA)
ASTM (USA)	=	American Society of Testing and Measurements (USA).
API	=	American Petroleum Institute (USA)
VDE	=	Association of German Electrical Engineers
IEC	=	International Electro-technical Commission
ISO-9000 QMS ISO-9001 QMS ISO-9004 QMS	=	**Quality Management Systems Standards** propagated by ISO, Geneva.
ISO-14001	=	Environmental Management System Standards;
IEEE	=	Institute of Electrical and Electronics Engineers (USA/ CANADA)
CEN	=	European Committee for Standards
NEMA	=	National Electrical Manufacturers Associations (USA)
FAO	=	Food and Agricultural Organization (of the UN)
WHO	=	World Health Organization (of the UN)
FPO	=	Standards Specified by the Ministry of Agriculture(India)
IND/P&T	=	Standards Specified by the Ministry of Communication
IND/ PHARMA	=	Standard Specified by the Ministry of Health (India)
IRS	=	Standards Specified by Research, Design and Standards Organization (RDSO) of the Indian Railways
IRC	=	Standards Specified by Indian Road Congress
JSS	=	Standards Specified by the Ministry of Defence(India) for defence production and defence/military supplies/ procurement.

NOTE: (1) Besides the above, there are National Standards Organizations in various countries which are members of ISO (Geneva).

- **ADDRESS OF ISO =**
  INTERNATIONAL STANDARDS ORGANISATION,
  CENTRAL SECRETARIAT,
  POST BOX-56, CH-1211,

GENEVA-20, SWITZARLAND.

- **ADDRESS OF BIS =**
  BUREAU OF INDIAN STANDARDS,
  MANAK BHAVAN,
  9, BAHADURSHAH ZAFAR MARG,
  NEW DELHI – 110 002.

NOTE: Refer to 'Directory of BIS' (Published every year) for:

(i) ***Listing of all Indian Standards;***

(ii) ***Listing of National Standards;***

(iii) ***Other relevant information;***

14.3 ENVIRONMENTAL MANAGEMENT SYSTEM (EMS) STANDARDS

An International effort to fight against ecological degradation, over exploitation of natural resources and to regulate Industrial Pollution:

REMARK: It is now imperative for Engineers to be aware of Environmental Issues and related prescribed standards.

A. With the growing population on the earth over the centuries, there have been ever-increasing exploitation of natural resources to meet the growing demands for consumption as well as for enhancing the standard of living. The rapid rate of industrialization and accelerated agricultural activities in the last two centuries resulted in over-exploitation of the natural resources so much so that not only the natural resources are getting rapidly depleted but such over-exploitation is also adversely affecting the environment to the detriment of the eco-system and the mankind. Realizing this, individuals and the societies in various parts of the world have started actively thinking to adopt measures to conserve and preserve the environment and stop/minimize the ecological degradation.

B. The issues like global warming, emission of green house gases, depletion of ozone layer in the upper atmosphere, depletion of green cover, rising of sea level, shrinking of glaciers, acid rain, increased concentration of harmful gases in the Atmosphere, pollution of surface and underground water sources and also land pollution are some of the burning issues faced by the modern world.

C. Most of the inputs of modern industries are non-renewable natural resources which are fast depleting. The major energy sources like coal, natural gas and oil are being consumed at an accelerated rate and may not be available in sufficient quantities for use by our future generations.

Therefore " Sustainable Development" is the buzz word now. The importance and urgency to conserve the natural resources and to reduce environmental degradation are now being realized the world over and efforts are on to minimize adverse impacts.

D. In view of the above, enlightened organizations world over are adopting environmental management systems and related standards to control impacts of their activities and products on the environment and to conserve valuable natural resources. The whole business philosophy is now changing . Now an organization is judged not only by its financial performance but also by its environmental performance and positive contributions to the society at large.

E. **Evolution of ISO-14000 series of Standards For Environmental Management**:

(i) The genesis of ISO-14000 series can be traced to the United Nation Conference on Human Environment in Stockholm in 1972. In this conference, more than113 countries were represented by more than 1000 delegates from 350 organizations. This conference resulted in a global action plan for the environment in which United Nations Environment Programme (UNEP) headed by Mr. Harlem Brundtland was formed. In 1987, the Brundtland Commission published a report titled "Our Common Future" calling for industry to develop effective Environmental Management Systems.

(ii) By the end of 1988, more than 50 world leaders had publicly supported the report. In 1989, the UN decided to convene the United Nations Conference on Environment and Development (UNCED), generally known as the **'Earth Summit'**. This Conference was held in Rio-de-Janeiro, Brazil, in June 1992.

(iii) In August 1991, ISO and IEC established the Strategic Advisory Group on the Environment (SAGE) to make recommendations regarding international standards for the environment. SAGE studied the British Standards Institution(BSI) standard BS7750 and other starting points for an ISO version.

(iv) In January 1993 , ISO created the Technical Committee 207 with the responsibility to develop uniform international standards on Environment Management Systems (EMS).

(v) Realizing the need for a uniform and systematic approach towards environmental quality attributes, the International Organization for Standardization (ISO) has produced since 1996, a series of International Standards on EMS which include following standards (Already Issued):

(vi) ISO-14000 FAMILY OF STANDARDS:

ISO 14001 : 1996	–	Environmental Management Systems Specifications with guidance for use.
*ISO 14004 : 1996	–	Environmental Management Systems : General Guidelines on principles of systems and supporting techniques.
ISO 14010 : 1996	–	Guidelines for Environmental Auditing-General Principles.
ISO 14011 : 1996	–	Guidelines for environmental Auditing-Auditing procedures.
ISO 14012 : 1996	–	Guidelines for environmental auditing-Qualification criteria for environmental auditors.
ISO 14015 : 2001	–	Environmental assessment of sites and organizations (EASO).
ISO 14020 : 2000	–	Environmental labels and declarations-General principles.
ISO 14021 : 1999	–	Environmental labels and declarations-self declared environmental claims (Type II environmental labeling).
ISO 14024 : 1999	–	Environmental labels and declarations-Type I environmental labeling Principles and procedures.
ISO 14031 : 1999	–	Environmental management- environmental performance evaluation Guidelines.
ISO 14040 : 1997	–	Environmental management-Life cycle assessment –Principles and frame work.
ISO 14041 : 1998	–	Environmental management-Life cycle assessment-Goal and scope definition and inventory analysis.
ISO 14042 : 2000	–	Environmental management-Life cycle assessment –Life cycle impact assessment.
ISO 14043 : 2000	–	Environmental management-Life cycle assessment –Life cycle interpretation.
ISO 14050 : 2002	–	Environmental management-Life cycle assessment –Examples of application of ISO standards.
ISO 19011 : 2002	–	Guidelines for quality and/or environmental management systems auditing (*This standard replaces ISO 14010,ISO14011 & ISO14012).

(F) **International Scenario:** EMS is gradually becoming a condition of doing business in many markets and regions. Some governments may require ISO 14001 Certification from suppliers of specific goods and services.

(G) **Indian Scenario:** ISO 14001 Certification has started taking roots in Indian Industries; Presently over 300 companies which include Chemical, Engineering, Petroleum, Processed Food and Hotels are working towards establishing and implementing ISO-14001 Standards, Presently, over 150 industries have already implemented ISO-14001.

(H) Environmental issues are becoming an important part of business planning. We must analyze and find what on earth can sustain and renew itself for life to go on. The future of mankind lies in a cleaner environment. It is duty of all organizations to work towards environmental conservation and resource conservation. Environmental Management is increasingly being accepted as a part of good business management strategy. If an organization has a sound environmental management system, it is certified against ISO 14001; apart from contributing to environmental conservation and resource conservation and control of pollution, the organization directly and indirectly gains through the following :

1. Provides a structured frame work for environmental management system.
2. Recognition for organization.
3. Increased business opportunities.
4. Better regulatory compliance.
5. Healthy working conditions.
6. Benefits arising due to lesser accidents.
7. Gets confidence of Insurers and financiers.
8. Good will in the society.
9. Good will among employees.
10. Economic benefits arising out of efficient use of natural resources.

■ ■ ■

15

Trouble Shooting for Mechanical Working Systems in Industrial Units

"To ensure satisfactory working of a system, you must also be conscious of the the circumstances that may go wrong in its operation and be prepared for relevant corrective steps, at the right time and, in the right measure."

–Author's View

REMARKS:

(i) Since there are a very large variety of Mechanical Working Systems (structures, machines, plants, equipment, devices) deployed in Industrial Enterprises, dealing with the processes of trouble-shooting (as part of Regular Maintenance and Servicing) for all of them will be a voluminous task, and therefore, the same cannot be covered in one chapter like the present one.

(ii) For the benefit of the young mechanical engineers, aspiring to make a career with Engineering Goods Manufacturing Enterprises, a limited coverage on machines, equipment, devices that are commonly deployed there, is being presented in the paragraphs that follow. The idea is to present an Awareness Module on the topic that may help familiarization with the tasks involved; a reasonable level of proficiency will come with practice over a period of time.

(iii) Many of the young mechanical engineers , in the course of their industrial career, may be engaged in the area of Plant Maintenance and Repair Services; the knowledge inputs provided herein , are expected to help them in discharging their tasks with the desired level of competence. Maintenance Engineers' performance is quite important and sometimes critical in the satisfactory and smooth running of the industrial enterprises.

15.0 MECHANICAL WORKING SYSTEMS COMMONLY DEPLOYED IN INDUSTRIAL UNITS

- **The Systems that operate under**:
  - Mechanical and thermal stresses arising out of loading, both in static and dynamic working conditions.
  - High temperature and high pressure operating conditions,
  - Subjected to external and internal forces,
  - Subjected to corrosive and detrimental working environment,
    - may develop certain malfunctioning/defects which may even lead to break-downs during the operating cycles—both regular and intermittent, due to wear and tear on body-parts, linkage mechanisms, and auxiliaries, which also gets aggravated due to aging and working over a period of time as per the recommended duty-cycles which may be based on the operating conditions such as :continuous duty, single shift, two-shift, three-shift working, to be considered with the level of machine-loading percentage.
- **Keeping the above (REMARKS and para-15.0) in view,** we are taking up, in the paragraphs below, **selected cases** of likely **trouble shooting needs/points** pertaining to the following mechanical working systems commonly deployed in industrial enterprises (**a Representative Listing):**

REMARKS: The Objective Purpose of Trouble Shooting can be stated as under:

(i) It presumes that any mechanical working system when ***subjected to continuous/regular, or even intermittent operations /usages, can develop malfunctioning/defects needing rectification to restore the system into healthy/satisfactory working condition;***

(ii) Trouble-Shooting efforts aim at identification and effectively tackling such malfunctioning/defects for eventual restoration to healthy working condition.

(iii) ***Engineers and Technicians assigned to do such jobs are supposed to be reasonably conversant with the working of the system as well as with the technological processes involved in carrying out trouble-shooting satisfactorily.***

While doing so, engineers may have to , situation demanding, think in terms of innovation to improve the post-repair functioning of the system.

15.1 MECHANICAL STRUCTURES IN INDUSTRIAL UNITS (excluding building structures which fall in the domain of Civil Engineering activities).

15.1.1 Static Structures

(A) These structures may be independently self-standing, or can be movable supporting parts or body parts of the main plants and machinery. While some of them can be stationary, others may be mobile, even rotating. These structural elements in the engineering industries are predominantly made of structural steels, alloy steels, grey iron castings, steel castings, steel forgings. In some cases, these structural elements can also be made of other non-ferrous metals like aluminum, brass, copper and their alloys, preferred because of certain valid engineering and technological reasons.

(B) Even in the stationary conditions, these may be subjected to certain internal and/or external forces/ stresses, vibrations, environmental corrosions , aging due to long period of service/usage.

(C) The Trouble-Shooting points that may require attention:

(i) Due to Aging: Rusting, corrosions, needing appropriate surface treatment with proper grade of painting/coating;

(ii) Deterioration in Foundation –needing repairs, re-laying, depending on the extent of damage;

(iii) Structural distortion/misalignment needing corrective actions which may include replacement of affected elements;

(iv) Accidental damage needing appropriate repairs;

15.1.2 Moving, Sliding, Rotating Structures

These dynamic structural elements, in most cases, will be of the nature of supporting/auxiliary sub-systems for a larger plant or machine, and are prone to wear and tear, and aging over a period of operation.

***The Trouble-Shooting points that may require attention:**

Loss of rigidity and distortion in the foundation, Geometrical configuration of the machine body, leading to misalignment effecting productivity and quality of output.

(i) Wear and tear on bearing points, hinges, fulcrums, suspension points;

(ii) Health of bearings, pins, spindle shafts,

(iii) Wear and tear on sliding surfaces/parts;

(iv) Malfunctioning of lubrication system(pumps, piping, drive controls);

(v) Malfunctioning of cooling and coolant supply system (pumps, piping, drive controls).

***Computerized system of Maintenance Scheduling and Record Keeping are recommended for satisfactory working.**

15.2 MECHANICAL PLANTS AND MACHINERY (Commonly Deployed in Manufacturing Industries)

15.2.1 Machine Tools

(1) In a manufacturing unit, a variety of Machine Tools are deployed to perform/undergo the technological processes as described in **Chapter-4, Chapter-5 and Chapter-8.** Mechanical Engineers must get reasonably familiarized with these aspects in order to be competent experts on carrying out the tasks of maintenance and servicing (including trouble shooting tasks).

It is also advisable that they get conversant with issues dealt with in the topics on **5S System and 'JISHU-HOZEN' work-culture** described in **Chapter-13.**

(2) Experience shows that generally, most plants/machines//equipment, on most occasions, somewhat like human beings, give out distress signals (on malfunctioning/defects) which an experienced (and sensible) engineer/technician can understand and interpret reasonably correctly, in order to diagnose the ailments and take remedial measures in time before it is too late (which may lead to break- downs).

Maintenance Engineers are supposed to be in regular touch with machine operators and user-area engineers to get timely feed back on any symptoms of malfunctioning/defects so that timely remedial measures can be taken.

(3) Organizing Preventive Maintenance:

(i) Trouble shooting efforts must be logically co-related with Preventive Maintenance efforts which help reduce frequency of occurrence of malfunctioning and break-downs (hence, down-time)..

(ii) Preventive Maintenance efforts need to take care of the following:

*Maintain a library/records of all machine tools installed (operation and maintenance manuals, other related documents duly catalogued and marked with shop-wise layout numbers/codes for **ease of identification and traceability);**

*Records of all previous repairs, preventive maintenance carried out;

*Draw-up annual/periodical preventive maintenance/servicing schedules and keep the concerned apprised of;

*Keep in stock, spares and other consumables in the coded bins and with identification tags;

*Earmark and line-up engineers, technicians and other helping hands as per schedules drawn up.

(4) Trouble-shooting efforts on machine-tools (Metal Cutting) should generally focus on the following:

(i) Condition of the Foundation:

*Settlement of machine bed, cracks developing, wedges displaced; corrosive environment causing pitting/erosions;

*unsatisfactory workmanship on foundation leading to unbalanced machine body and loss of rigidity;

(ii) Excessive vibration beyond permissible limits resulting in:

*Machine getting loose on bed, anchoring bolts loosened, wedges disturbed/displaced;

*Machine loading eccentric, overloading, bearings worn out/ broken, bearing housing developing cracks; spindles out of alignments and distorted/wore off beyond acceptable limits ,

*Drive motor developing malfunctioning (electrical faults, bearings failure);

*Job-holding and tool-holding fixtures/devices developing defects/distortions, improper clamping (distortions in clamp fixtures);

(iii) Lubrication System developing malfunctioning:

- Pumps not working properly(propeller eroded, propeller vanes developed cracks),
- Pump motor failure, control system failure;
- Wrong grade of lubricant used, -
- Failure/leakage in the piping/hoses; lub oil very old and dirty and filters choked;
- Electrical power supply system malfunctioning;
- Sequence of operation of cyclic lubrication got erratic;

(iv) Coolant System malfunctioning:

- Points similar to those mentioned at (ii) above;

(v) Drives, associated controls and instrumentation:

- Base-plate/bed and anchoring bolts loose on foundation;
- Coupling bolts on drive shaft-flanges worn-out/loosened; drive shaft developed misalignment (eccentric loading);
- Electrical cable faults, relays developing malfunctioning;;
- Bearing failure/lack of lubrication on the drive motors shaft;
- Under-voltage condition in electrical power supply;
- Brush-gear on motor slip rings/commutator (in case of DC motor, AC Slip ring motors, AC Commutator motors);
- Driving chains and sprockets worn-out and causing erratic movement.

- Indicating instruments not working properly (erratic);
- Drive motor subjected to over-loading;
- Ambient temperature too high;
- Power transmission system (gear-train, clutches) developing malfunctioning leading to unbalanced operating condition:
- Gear teeth broken/eroded, loose on meshing, wear and tear on Lead Screws causing loss of precision in movements, inadequate lubrication, low on gear oil content.

(vi) Hydraulic and Pneumatic Systems developing defects:

- Pumps not working to capacity due to worn-out/damaged propeller/impellor;
- Cylinders eroded inside causing loss of pressure in the system;
- Plunger/pistons and seals are corroded/damaged leading to fluid leakages and loss of pressure in the system;
- Electrical faults in the drive motors and associated controls; under-voltage in power supply ;
- Bearings in the pump and pump motor worn-out/damaged; lubrication failure/inadequate flow causing bearing failures;
- Hoses and pipes are damaged and leaking causing pressure drops in the system;
- Actuators not working properly(defective plunger, springs, washers, fixtures);
- Inadequate pressure in the compressed air pipe lines; excessive moisture in the pipelines;
- Air compressors not working properly causing pressure drops in piping the system (check drive motor performance, electrical faults, bearing failure; lubrication failure, control system faults);

(vii) Controls and Instrumentation developing malfunctioning:

- Control cable network developing defects due to mechanical damages, electrical faults/insulation failures,
- Erratic functioning of relays due to erratic input system voltage;
- Electronic circuitry and circuit-elements (cables, relays, actuators etc.) developing faults;
- Instruments developing defects, losing accuracy; recalibration and replacement may be required;
- Display panels (electrical-electronic elements) may develop defects;
- Servo-drive controls system may develop malfunctioning (electrical and electronic elements defective, mini and micro

motors failures, linkage mechanisms not working properly due to aging/wear and tear);

- Ambient temperature and humidity too high causing erratic functioning of electro-mechanical-electronic system elements;

(viii) CNC Controls (wherever incorporated) (*):

Trouble shooting may relate to all or some of the following:

- Computer programmes not compatible, erratic commands,
- Selected software not usable/suitable;
- Data entry cards/tapes, CDs developing defects due to prolonged usage,
- Malfunctioning in the on-board computer; network disruption between central station and local work-station;
- Job-setting and tool-setting not properly done causing machining distortions on the work-piece;
- Encoder/Decoder systems not functioning properly, not compatible with the computer being used; feedback signal transmission system not functioning properly;
- Servo-drive and servo-controls not functioning properly;
- Linkage mechanisms developing snags(worn out, displaced);
- Ball Screws and roller chains developed defects due to wear and tear, causing erratic movements of critical machine-elements, thus affecting accuracy and quality of outputs;
- Tools being used are not appropriate for the work-piece (wrong size and cutting geometry, blunted cutting edges);

(*)REMARKS: Trouble Shooting on CNC Hardware (electronics) system is a specialized job normally to be handled by trained Electronics Engineers/Technicians; mechanical engineers are not supposed to fiddle with the same, but they should be aware of such requirements and take help from the concerned experts as and when such exigency arises.

(ix) Cutting Tools used with Metal Cutting Machine Tools:

REMARK: Readers are advised to read the topic presented at Paragraph 8.4 in Chapter 8 before reading the following statements.

Successful usage and productivity of machine tools, to a large extent, depend on the **quality and compatibility of** the cutting tools being used; therefore, it is essential to **take care of the following issues:**

*A good level of knowledge regarding the working conditions required for various types of tools (configurations, matching sizes, material of construction, cutting- edge geometry, cutting angles/rake angles, tool-holding arrangements, work-piece holding fixtures to this is a specialized job and require the services of trained and experienced tool room engineers/technicians;

- Over a period of usage, cutting edges of tools become blunt (and occasionally broken also) needing re-sharpening, and in extreme cases , need replacement with new sets.
- Lack of proper coolant flow at the cutting points;
- Chips/swarfs getting entangled around cutting tools creating hindrance and undesirable scratches on work-piece;
- On being informed/apprised by the operator/shop engineer, the tool engineer/production technologist are supposed to look in to the problem and help in remedial steps (which may include recommendation for replacement).
- Under normal operating conditions, trouble shooting activities may be limited to re-sharpening/re-grinding and certain minor re-adjustments in the tool body/shank portion.
- Jigs and fixtures and blunted cutting tools can be repaired /re-conditioned/re-sharpened, provided required facilities and trained personnel are available in-house; in case facilities and trained personnel are not available in-house, services of external agencies may have to be hired.
- Incase of major damage beyond economic repair/reconditioning, better be discarded and new ones put to use.

(5) In the case of Metal Forming Machine Tools:

- These generally fall in the category of metallurgical/foundry equipment such as Hydraulic and Pneumatic Forging Hammers, Presses, Hot and Cold Extrusion plants and the like.
- **Trouble Shooting issues** may generally relate to:
  - Those listed under the paragraphs (4)(iv, v, vi) above.
  - Defects/distortions developing on forging and pressing dies and their fixtures;
  - Work-piece heating furnaces not working as per prescribed parameters of heat and temperature;

(6) In the case of Metal Working Machine Tools:

(A) These generally refer to:

(i) Power Presses(crank type) fitted with Electro-Mechanical – Pneumatic Operating systems,

(ii) Hydraulic Presses operated by Electro-Hydraulic systems

(iii) Plate and Bar Straitening Presses working with electro-hydraulic and electro-magnetic operating systems.

(iv) Shearing and Cropping machines operated either by electro-mechanical(power crank presses) , or, electro-hydraulic operating system.

(B) **Operating troubles that may be encountered**:

(i) **In the case of Machines operating with Electro-Mechanical-Pneumatic systems**:

- Low compressed air pressure in the supply lines (pipelines/ hoses ruptured, joints and control valves leaking);
- Actuators defective,
- Pneumatic clutches and brakes not working properly;
- Electrical faults on drive motor and associated controls;
- Failure of bearings/bushes (worn out/broken, lack of lubrication);
- Flywheel bushes worn out, V-belts eroded/snapped/slipping;
- Roller-table not working properly, sprockets and chains worn out resulting in erratic/sluggish movements;
- Gear box developed defects(worn out /broken gear teeth, bearing failures, gear oil level low , gear oil too old and dirty/.
- Machine foundation developed cracks, anchoring bolts loose;

(ii) **In the case of Machines operating with Electro-Mechanical-Hydraulic Systems:**

- Operating problems by and large will be similar to those listed above under paragraphs (4)(v,vi);
- Linkage mechanisms, gears, pinions, sprockets, chains developing malfunctioning due to wear and tear, breakage of teeth;
- Bearing/bush failures, misalignment of drive shafts, failure of coupling bolts/clutches;
- Cutting edges/knife edges blunted/broken;
- Eccentric loading due to faulty tools, jigs, fixtures;
- Indexing on compound dies not properly adjusted;

15.2.2. Pumps

(1) **Readers are advised to read paragraph 10.1.1 in Chapter-10 for a general familiarization on pumps**.

(2) Pumps have either reciprocating(sliding) parts or rotating parts depending on the types o pumps used.

(3) **The Pumps generally may experience the following kinds of operating troubles which the engineers will have to deal with:**

(a) **Excessive vibration due to:**

– Misalignment of pump body/ drive shaft/ loose foundation bolts;

- Propeller developed dynamic unbalance due to damage/erosion in blades;
- Bearing failures;
- Malfunctioning of diving prime-mover(bearing failures, engine failure/fuel injection problem, electrical motor failure/electrical faults/under-voltage supply);
- Bearing seals leaking/ gaskets ruptured , plumbing problems on intake side and delivery side piping;
- Discharge rate and head low (under capacity, plumbing defects);
- Pump body getting over-heated due to lack /no-flow of intake fluid (supply-side problems); air-locking on intake side;
- Excessive noise due to propeller unbalance, bearing jamming/ lack of lubrication;

(b) Piston/Reciprocating Pump:

- Short-stroke due to defects in piston rod/piston cylinder, piston ring breakage/jamming;
- Seals leakage, slides eroded;

(c) Gear Pump:

- Gear teeth eroded/ broken,
- Out of alignment/meshing due to axle/shaft bearing failure;
- Priming problem at start{air locking/leakage, intake side dry/ no fluid, suction- head fluid level too low;
- Pump under-capacity for the desired duty–cycle/fluid loads;

(d) Multi-Stage Pumps:

- Propellers misaligned on the common drive shaft;
- Inter-stage pump-body gasket leaking;

15.2.3 Air/Gas Compressors

(1) Readers are advised to read the brief on Air Compressors presented in paragraph 10.1.2. in Chapter-10.

(2) Operating problems for trouble shooting are more or less similar to those stated above under paragraph 15.2.2(3).

- **Additionally:**
  - Malfunctioning of non-return valves(springs and plunger defective/broken/eroded, gaskets leaking, plunger seat eroded);
  - Receiver/storage tank developing leakage, piping/plumbing failures;
  - Pressure drops in the pipe lines due to excessive demand/ leakages;

15.2.4 Power Presses

(1) Readers are advised to read the brief on Power Presses presented at paragraph 10.1.3. in Chapter-10.

(2) Operating problems that are commonly encountered can be listed as under:

(a) In the case of Electro-Mechanical (Crank) type:

- Machine foundation weakened, loss of rigidity of machine body;
- Drive motor developing electrical faults, under voltage supply;
- Crank-shaft bearings/bushes eroded causing erratic/inadequate pressing power delivery;
- Defects on the dies being used(eroded, distorted,);
- Inadequate lubrication in the bearings/bushes;
- Electrical faults in the control pendant ;
- Malfunctioning in the pneumatic clutch (where used);
- Fly-wheel unbalanced due to bearing/bush wear and tear;

(b) In the case of Electro-Hydraulic and Electro-Pneumatic Operating System based Presses:

> Operating problems by and large will be similar to those listed above under paragraphs -15.2.1(4)(iv, v, vi);

- And also problems similar as listed above at para-15.2.4(2)(a).

15.2.5 Pneumatic Hammers

Referring to paragraph-10.1.4 in Chapter-10, the operating problems that may be encountered will be by and large similar to those listed above under paragraph-15.2.1(4)(iv, v, vi);

- **Additionally:**
  - Foundation developing defects (un-even settlement, anchoring bolts loosened);
  - Hammer body-frame developing cracks/distortion;
  - Dies developing distortions/erosions;
  - Rams erodes, ram seals broken, ram misaligned/broken.

15.3 FURNACES AND OVENS

>>Readers are advised to read the brief on the topic presented in Paragraph-10.1.5 in Chapter-10.

REMARK: Furnaces for melting metals are excluded from this discussion as these are supposed to be covered under Metallurgical Engineering/ Foundry Practices.

15.3.1 Furnaces

There are Three Broad types of Furnaces commonly in use in Industrial Units as being presented below.

>> Operating problems that are generally encountered for different types of Furnaces can be listed as under:

(1) In the case of Oil Fired Furnaces:

(a) **Not achieving required heat and temperature parameters due to:**
- Inferior grade of furnace oil being used (poorer calorific value);
- Excessive heat leakage in the furnace chamber, heat insulation degraded/fallen off;

(b) **Heat load is more** than the designed capacity;

(c) Defective burners(clogged, nozzles eroded/distorted/melted);

(d) Air supply not adequate, fans/blowers not working to capacity; fans/blowers developed defects due to aging;

(e) PID Controller not functioning properly; control wirings faulty;

(f) Oil pumps not functioning properly, electrical faults on drive motor; bearing failure on pump motor;

(g) Pump propeller vanes eroded/distorted/broken;

(h) Pre-heating devices for oil and air not functioning properly;

(i) Low draught in the exhaust/stack(creating back-draught and un-even heat circulation);

(2) In the case of Gas Fired Furnaces:

Operating problems are by and large similar to the ones listed at (1) above except that furnace oil is replaced by fuel gas.

***Additionally:**

(a) Gas fired furnaces are lees efficient since fuel quality(calorific value, supply volume and pressure etc.) are not always consistent; a constant control on Air-Gas mixture injection is to be ensured; also more numbers of Burners are to be placed to achieve the desired level of heat and temperature regimes.

(b) For the same rate of heat and temperature ranges, volumetric size of a gas fired furnace chamber will be bigger which brings in the problem of:
- Maintaining even distribution heat and temperature in the different zones of the fire chamber;
- Heat leakage level through the body and doors may be higher;
- Chances of collapse of refractory lining increases over a period.

(3) In the case of Electrically Heated Furnaces:

There are generally Four Sub-types as under; operating problems that can be encountered for each of them:

(3-1) Electrical Resistance Heating Furnaces:

- Low heat and temperature due to inadequate numbers of heating elements; frequent burn-outs due to low quality of elements; heat-load is higher than recommended;
- Lower electrical supply voltage;
- Resistance elements more in the circuit causing lower current flow;
- Leakage of heat due to inadequate insulation/refractory-lining/ collapse of refractory lining;
- Defective instruments giving wrong readings/indications;
- Malfunctioning of control apparatus, defects in heat-sensors/ disconnection/melt-out of **thermo-couples;**
- Electrical cables are failing due to bad heat–resistant insulation quality;
- Problems with fans and blowers(drive motors malfunctioning, electrical faults, bearing failures, fan leaves damaged/distorted);

(3-2) Electric Arc Furnaces:

Since these furnaces are of unique types and are pre-dominantly used for melting of metals in Metallurgical Industries /Foundries, they are supposed to be covered under Metallurgical Engineering Fields/Foundry Practices; therefore not being dealt here in this book.

(3-3) Electrical Induction Furnaces:

(a) **Malfunctioning of the MG Set due to** :
- Electrical faults/insulation failure, bearing failures,
- Excessive vibration due to weak/loose foundation/coupling bolts loosened/ eroded/broken;
- Single-phasing of drive motor; low supply voltage causing lower motor rpm and lower induction frequency generation;

(b) **Induction Generator** (normally, asynchronous type) developing electrical faults, lower speed of the drive motor due to low supply voltage;

(c) Problems in cooling water supply(pump failure, pipes/hoses developing leakages/ water intake insufficient);

(d) Malfunctioning of control apparatus and frequency converter and stabilizer;

(3-4) Electric Salt-Bath Furnace:

These are basically Heat-Treatment furnaces used in Tool Rooms for hardening HSS Tools, fasteners of special grades, cams, levers, small heavy-duty gears sprockets, rollers, inner-races of bearings and the like. The electrodes, submerged in the salt baths , create a continuous Electrical arc.

The following operational problems may be encountered:

- Electrical problems: low supply voltage, cable faults, disruption in arc-formation and continuity ;
- Electrodes burning out /eroding at higher rate than prescribed; (bad quality),Electrodes not positioned properly;
- Refractory collapsing early/frequently due to poor quality, over-heating(Arc regulation not proper);
- Loading-unloading mechanisms not working properly(linkages/ chains /pulleys developed defects due to wear and tear);
- Control apparatus elements and instruments not working properly;

15.3.2 Ovens

Ovens, normally smaller than furnaces (can be also in desk-top versions), are used for heating, drying, baking and curing, both for metallic and non-metallic work-pieces; the heating source can be electrical resistance heaters, oil fired burners, gas (LPG, Producer gas) burners; most of them are designed for limited duty-cycle. **Operating problems are likely to be similar to those listed above for furnaces, but in a smaller scale.**

15.4 BOILERS

REMARK: Readers are advised to read the brief presented in Paragraph-10.1.6 in Chapter-10 in order to Issues presented below.

15.4.1 Unfired Boilers

These are commonly configured in two versions as under:

(1) Heat-Exchanger Type Boilers:

Operating problems can be as listed under:

- Inadequate heating available from the primary source;
- Heat leakage in the system: inadequate insulation ,both in the body as well as in the pipe-lines;
- Clogging in the tubes and headers due to scaling/crusting;
- Fluid leakage from the tube plates placed inside the barrel;
- Ruptures in tubes, failure of gaskets in the flanges;
- Failure of control valves, non-return valves, piping rupture, plumbing defects, piping hangers failing ;
- Excessive loading in the secondary circuit;
- Instrumentation and control apparatus malfunctioning;

(2) **Electrically (Resistance) Heated Boilers:**

These are normally small size ones to generate steam for limited usage in laboratories, hospitals, hotels, residential units.

Operational problems experienced can be as listed under:

- Electrical faults, inadequate power input (low voltage, heating elements inadequate/burn-outs, cable faults);
- Insulation failure: both thermal and electrical;
- Leakage/rupture in water tubes and pipes, failure of gaskets in flanges;
- Intake water flow is low in quantum and pressure;
- Failure of thermostats leading to over heating;
- Defects in steam blow off valve;
- Instruments not working properly, erratic/wrong indication;
- Overloading due to prolonged operation/usage;

15.4.2 Fired Boilers

REMARKS:

(i) Readers are advised to read the briefs presented in paragraph -10.1.6(2). in Chapter-10.

(ii) This presentation is limited to Steam Generation Boilers using treated(de-mineralized) water;

(iii) These boilers are fired by burning (in their fire chambers)fossil fuels such as Furnace Oil, coal, lignite, Natural Gas;

>>Operational problems that are generally encountered can be listed as under (*assuming that the Design/Engineering and construction of the boiler have followed the standard/prescribed and proven technological systems and quality regimes*):

(i) Clogging of tubes/pipes and headers due to scaling; higher levels of ingress of mud/suspended particles into the cooling and DM Water supply systems;

(ii) Rupturing of tubes/pipes; leakage due to failure of gaskets, valves;

(iii) Failure of thermal insulation;

(iv) **Low heat due to:**

- Inadequate level of fuel supply to the burners in fire chamber;
- Inferior quality of fuel(lower calorific value, contaminated fuel);
- Defective burners/clogging of burners,
- Defective /malfunctioning of fuel transportation and delivery systems affecting flow of fuel supply,

(v) Leakage in de-aerator system, HP and LP Heaters, Hot well pipes,

(vi) Malfunctioning/failure of steam flow control valves,

(vii) Malfunctioning of primary water supply system(failure of pumps and pump motors, electrical faults, rupture in pipes, failure of gaskets in flanges, intake water(suction side) level low, head-losses on the delivery side).

(viii) Malfunctioning of DM Water supply system (water treatment plant not working properly, leakage in pipe lines, failure of pump and pump motor and associated control gears,

(ix) Malfunctioning of feed water supply system : electrical faults in drive motors, propellers eroded/ distorted due to aging, leakage in pipes, control valves developed defects/leakages, spindles and gates jamming;

(x) **Low steam temperature and pressure due to:**

- Inefficient working of fire chamber equipment;
- Points mentioned at (iv) above;
- Higher level of heat leakage from boiler body and steam pipelines;
- Steam leakages through pipe flanges and control valves ;
- Steam pipe lines are too long and too many bends on their run path;

(xi) Malfunctioning of cooling towers: electrical faults and bearing failure pump and fan motors, plumbing defects, fan blades eroded/damaged, failure of gratings inside tower due to aging, ambient temperature too high effecting cooling rate;

(xii) Deficiencies in fuel storage and delivery systems; snags developing in conveyor system (drive motor failure, belting eroded/snapped, rollers eroded and their bearings failing, structural failures; controls apparatus developing snags);

(xiii) Defects developing in ash and slag handling/disposal system (mud pumps and drives not functioning properly (electrical faults, pump propellers eroded/damaged,

(xiv) Super-heaters developing leakages due to tube failures (clogging, tube rupture);

(xv) Malfunctioning of ID and FD Fans (electrical faults in drive motors and associated controls, bearing failures, fan blades damaged/ distorted);

(xvi) Defects/operational problems developing in Station Sub-station: switch-gear failures, transformer failures, over-loading, cable faults;

(xvii) Problems developing on Boiler Drum: thermal insulation and cladding collapsing at some points, steam pipe line connecting flange-joint gaskets failing**; in rare cases**, supporting structures showing signs of deterioration/distortions;

(xviii) Bending/distortions developing in water tubes in the fire walls inside the fire chamber;

(xviii) Malfunctioning developing in the control panels in the Central Control Room (failures of electrical, electronics elements, instruments, cables, failures in computer networking and communication systems);

(xix) **Regular Attention is required for the following to keep them in healthy(and dependable) operational conditions:**

- Fire prevention and fire fighting systems and equipment;
- Safety measures and emergency services; Security Services;
- Emergency Power-backup System(UPS/Battery Banks);
- Ensure obtaining periodical (Statutory) Certification from the Boiler Inspector and Factory Inspector.

15.5 HEAT EXCHANGERS

(1) Heat Exchangers being used as Un-fired Boiler:

Operating problems encountered will be more or less similar to listed above at paragraph-15.4.1(1);

(2) Heat Exchanger being used as a Cooler of Hot Fluids flowing in the Primary Circuit:

Operating problems encountered will be more or less similar to those listed under:

- Leakages from tubes (ruptures) and flanged joints (gasket failure),
- Leakage from the tube plates;
- Scaling/clogging inside tubes
- Flow of cooling fluid is not enough to bring down temperature to the desired level (to check pump capacity/performance);
- Ambient temperature of cooling fluid is high (check cooling tower performance in case the cooling fluid is in a circulatory mode);
- Pump and motor may develop electrical faults and bearing failures;
- Temperature of the primary circuit fluid may be too high with respect to the designed capacity of the heat-exchanger;
- There may be defects in the receiver tank piping and the non-return valve;
- There may be failures (corrosion, distortions) in the supporting structural elements due to aging;

15.6 INDUSTRIAL FANS AND BLOWERS

Common operating problems experienced can be listed as under:

- Excessive vibration and noise: defective /dry worn out bearings, loose supports /brackets,
- Fan blades unbalanced/distorted;
- Electrical faults and bearing problems(lack of lubrication) in fan / blower motors, failure of V-Belts due to aging/wear and tear;
- Speed/rpm lower than required: supply voltage low, defective starter capacitor;
- Defects in controls/switch-gears;
- Placement of fans and blowers not correct;

15.7 MECHANIZED VENTILATION SYSTEM

- **Readers are advised to read the brief on the topic presented at paragraph-10.1.10 in Chapter-10.**
- **The operating problems generally encountered can be as under:**
  - **Fans and Blowers:**
    - Similar to the ones listed above at paragraph- 15.6.
  - **Ducting:**
    - Structural failures in hangers/brackets;
    - Ducting size is inadequate, too long and too many bends;
    - Grills choked, not functioning properly;
    - Improperly placed ingestion and discharge points for gas/ fumes/hot Air;
    - Corrosive fumes and abrasive dusts eating away internal lining; lining pealing away;
  - **Baffles** are jamming/eroded/loose and wrongly placed;

15.8 AIR CONDITIONING SYSTEMS

- **Readers are advised to read the brief on the topic presented in paragraph-10.1.12 in Chapter-10.**
- ***THE SCOPE OF DISCUSSION HERE IS LIMITED TO THE*** PREMISES (OFFICES, WORK-SHOPS, STORAGE BUILDINGS, LABORATORIES) OF THE INDUSTRIAL UNIT.

15.8.1 Air Conditioning In Low-Temperature Zones

(1) When the ambient temperature is below a comfortable level, or, is lower than the prescribed ambient temperature level in the premises engaged in certain scientific/technological processes, measures are taken to increase the prevailing temperature to the desired level.

This can be achieved by certain commonly adopted methods as described in the paragraphs -10.1.12(A)(1- 4) in Chapter 10.

(2) Each of these processes(as mentioned above) will have certain common as well as unique operating problems as listed out below:

(A) In the case of Space Heating by deploying Electrical Resistance Heating Devices:

- Heating elements burning out due to prolonged usage, electrical faults; insulation failure in the supports and cables;
- Blowers developing defects (electrical faults, bearing failure, fans blades eroded/distorted, low supply voltage);
- Inadequate numbers of heaters vis-à-vis the space to be heated;
- Abuse of heaters by users;
- Heat leakage/loss due to high level of ingress of cold air from outside due to inadequate sealing of premise;
- Malfunctioning of AHU (=Air Handling Unit); ducting leakage*;
- Discomfort due to lack of Humidity Level Control (device either not installed, or is not functioning as prescribed);

**NOTES:*

(i) The same ducting can be used during the summer months for chilled air circulation provided thermal insulation (inside/ outside) is of matching grade and quality.

(ii) In case the same ducts are used, AC(gas) Compressors and the chilling plants are to be effectively isolated from the system.

(B) In case of Space Heating is done by Generating Hot Air in Furnaces and Blowing and circulating the same through ducts running in the targeted premises:

- Since these furnaces, most likely, will use fossil fuels like furnace oil, Producer/natural gas, coal, **the operating problems** are likely to be similar to those listed above in the case of **Furnaces** at paragraphs 15.3.1(1,2) and **Fans and Blowers** at paragraph-15.6.

***Additionally:**

- Structural failures in ducts, deteriorated thermal insulation, clogging in baffles,
- Noise emanating from blowers (bearing failure, unbalanced fan blades);
- Problems in AHU and filter elements (clogging, ingress of smoke/ flue gas into the hot air ducting);
- Malfunctioning of ash and slag disposal system, associated environmental issues;

- Problems associated with fuel storage, fuel handling, environmental issues due to emissions of flue gases from furnaces;

(C) In case of Installing Boilers to generate Steam for circulation into the premises through insulated pipelines which in turn get connected to locally placed Heat Radiators:

This is a costly and complex installation and is normally employed in places where ambient temperature goes near sub-zero or below, during winter months.

Operating problems are more or less similar to those listed above at paragraph-15.4.2 (Fired Boilers).

(D) Using Surplus/discharge heat from process plants (located in the same factory premises) to generate Hot air to be circulated through ducts as in (B) above:

Operating problems that may be encountered:

- Consistency of availability of such surplus/discharge heat to the requires quantum at the required period;
- Usual problems with insulated pipelines and ducting as mentioned at (B)above;

(E) Heating by Harnessing Solar Rays:

This method is NOT normally employed for Space Heating because of Techno-economic and locational limitations; this method (using solar parabolic reflectors) is commonly adopted for generating hot water for supply to target areas through insulated pipes (installed on roof-tops in hotels, hospitals, residential units offices, labs).

Operating problems experience may be as under:

- Low heat during rainy and cloudy days;
- Reflectors get covered by dust needing frequent cleaning;
- Aging of thermal insulation needing repairs off and on;
- Structural failures during storms/cyclones and due to aging;

15.8.2 Summer Air Conditioning (Cooling)

- **Readers may refer to the brief presented at paragraph-10.1.12(B) in Chapter-10.**
- **Depending on the size of the Premises, the following three methods are generally employed to achieve cooling at the desired level. Each method is presented below along with their respective operating problems:**

(1) Window Type Air conditioners:

Operating problems that generally experienced:

- Noisy due to excessive vibration in the loose body structure, unbalanced compressor unit, unbalanced fan/blower, loose mounting of compressor unit;
- Electrical faults in compressor motor, bearing failure, capacitor failure,
- Electrical/electronic control system not functioning properly;
- Inadequate cooling due to gas leakage/tubes corroded/ruptured; inferior grade of coolant gas used;
- Inadequate cooling due to larger volume of the premise/under-capacity Ac unit installed, ingress hot air due bad sealing of the premise;
- Low supply voltage to compressor motor/failure of voltage stabilizer;

(2) Split Type Air conditioners:

*Operating problems, by and large, are similar to those listed at (1) above;

***Additionally:**

- Flexible insulated piping-ducting developing leakage, run length is either too long, or, too short;
- External fan/blower choked due to ingress of dust, fan motor not working properly(electrical faults, bearing/bush worn out);

(3) Packaged Type AC Units:

- **Readers are advised to refer to the brief on the topic presented** in paragraph-10.1.12 (B)(2).
- **Operating problems encountered can be listed as under:**

*Most of the problems are similar to those mentioned above at (1) and (2);

***Additionally:**

- AHU not functioning properly, air filters getting choked, fans/ blowers developing electrical faults,;
- Water supply system not working properly, tubes/hoses getting clogged/ruptures occurring,
- Thermostat switches defective/wrong setting causing frequent tripping of gas compressor,
- Depletion of refrigerant gas/brine solution due to leakage;
- Control apparatus malfunctioning/switches and relays not working properly, instrumentation failures;
- Over-loading of compressor due to continuous running for long hours;

(4) Central Refrigeration Type Plants (with associated Chilled-water Plant, AHU, Ducting):

- Readers are advised to refer to the brief on the topic presented In paragraph-10.1.12(B)(3);
- **Operating problems that are generally encountered can be** listed as under:
    - Malfunctioning of gas compressors, compressor motors, AHU and associated sub-systems, by and large, similar to those listed above at paragraphs-15.8.2 (1, 2, 3);
- **Other issues that may crop up:**
    - – Malfunctioning of chilling plant and associated auxiliaries; leakage and damages in the insulated pipe lines(due to aging),
    - – De-humidifier not working properly affecting control on humidity Level within prescribed level.
    - – Clogging in heat-exchanger, leakage in tubes and tube plates,
    - – Cooling towers not working properly(electrical faults in fan motors, fans blades eroded/distorted, gratings collapsed due to aging, sprinklers clogged, plumbing failures in water supply and return pipe lines causing cooling cycle imbalance;
    - – Malfunctioning of control apparatus(electrical/electronic elements and switch-gears not working properly due to aging/ insulation failures;
    - – Structural failures (mainly due to aging) of ducting, thermal insulation pealing off/inadequate, baffles not working properly;
    - – Sealing in premises not properly done resulting in ingress of warm air from outside causing increase in cooling load/rise in room temperature;
    - – Subsequent unplanned addition of premises area/volume causing overloading on cooling cycle and increase in room temperature;
    - – Inferior grade and quantum of refrigeration gas/fluid used;

15.9 REFRIGERATION PLANT

- **Readers are advised to refer to the brief on the topic presented at paragraph-10.1.13 in Chapter 10.**
- **Operating problems are by and large similar to those listed at paragraph-15.8.2 (Summer Air Conditioning).**
- **The following special features may be kept in view while planning trouble-shooting on the composite installation:**
    (i) The chilled fluid (normally water) will have a temperature range 5°C to 7°C, and therefore, the thermal insulation on the equipment and pipelines will have to be of matching grade with embedded

fixtures; with passage of time, there can be deterioration needing repairs/relaying;

(ii) There can be malfunctioning in the various auxiliaries including electrical faults and control system/instrumentation failures;

(iii) Rotating equipment (motors, fans etc.) will have wear and tear needing repairs and replacements from time to time; lubrication system need attention;

15.10 HUMIDITY CONTROL

- **Readers are advised to refer to the brief on the topic presented at paragraph-10.1.14 in Chapter-10.**
- **Depending on the type of Humidifier and De-Humidifier installed, operating problems have to be indentified and servicing and required repairs/replacement of parts/auxiliaries have to be carried out.**
  - All rotating equipment and stressed pipes/tubes will be prone to developing defects/wear and tear with passage of time and therefore will need preventive as well as break-down maintenance.
  - Engineers and technicians are supposed to get familiarized with the instructions/tips given in the Operating Manuals supplied by the manufacturers.

15.11 PIPE LINES OF MECHANICAL SERVICES

- **Referring to the brief presented a paragraph-10.1.15 in Chapter-10, certain pipelines in the factory premises will be identified under Mechanical Services pertaining to the transportation of fluids such as steam, hot water, chilled water, gases, oil, hot and cold air, compressed air, hydraulic fluids, and these pipelines serve/relate to certain Mechanical Plants and Machines/Equipment, and therefore, these are earmarked to be dealt with by the Mechanical Engineering Department.**
- **To facilitate proper identification and avoid mix up, all pipe lines are colour-coded as per standards prescribed;**
- **Problems that are generally experienced with pipe lines:**
  - Plumbing failures/mistakes (wrong grade of pipes used), corrosions and ruptures, gaskets failures at flanged joints, clogging/internal scaling;
  - Structural failures in supports/hangers/brackets; vibration and noise beyond permissible limits;
  - Failure of thermal insulation (wherever used);
  - Failure of control valves and actuators due to wear and tear/long-time usage;

- Pressure drops due to head-loss/ too many bends in the run, inappropriate sizing of pipes (diameter); malfunctioning of pumps/ pump-motors (speed and pressure drops);
- Discharge head is more than recommended for the pump used causing head loss/non-delivery;

15.12 HYDRAULIC AND PNEUMATIC OPERATING SYSTEMS

- **Readers are advised to refer to the brief on the topic presented at paragraphs-10.1.17 and 10.1.18 in Chapter-10.**
- **Operating problems experienced are by and large, similar to those listed ABOVE under paragraph-15.2.1(4-V).**

***Special attention needed to the following**:

- Malfunctioning of hydraulic pumps and pump motors;
- Malfunctioning air compressors and their dive motors;
- Failures in pipelines and hoses, defects in actuators and valves;,
- Wear and tear inside hydraulic cylinders, failure of seals
- Pressure loss in the system due to leakage and excess run and too many bends in the pipe lines/hoses;
- Too much moisture in the air line, moisture traps are not functioning Properly/ not cleaned and drained out periodically;

15.13 MATERIAL HANDLING EQUIPMENT

- **Readers are advised to refer to the brief on the topic presented in paragraph-10.1.7 in Chapter-10.**
- **Operating problems encountered with the types of Material Handling Equipment that are commonly deployed in Industrial Units, are listed out in the paragraphs that follow:**

(1) EOT Cranes:

- Drive motor controllers: electrical faults, hoist drum controller contacts wearing off, electrical faults in drive motor, motor bearing failures;
- Power supply trolley lines and current collectors developing snags;
- Due to wear and tear; supporting structures got misaligned/loosened;
- Steel ropes snared/damaged, hoist drum movement/speed un-even;
- Gear box developing snags: teeth breakage, axle/shaft bearing failure;
- Misaligned meshing; lubrication/gear oil inadequate/old and dirty;
- Steel tyre faces eroded beyond acceptable limits causing uneven movements/side-shifts;
- Steel tracks eroded and got misaligned causing uneven/side-shift movements;

- Braking system developing snags: electro-magnet coils burning out/ hydraulic thrusters failing to work properly, brake shoes eroded beyond limits: hydraulic pump not working properly, failing to deliver required thrust power/plunger rods getting obstructed due to seals damaged and oil leakage.
- **Crab trolley** movement tardy due to misalignment of cross rails,
- Gears/gear teeth breakage/bearing failures;
- Pendant controller developing electrical faults, contacts eroded;
- **Radio-Remote Controller** (if used) developing malfunctioning;
- Track-end buffers damaged due to: failure of limit switches, operator's mistake/carelessness;
- Lifting hooks developing snags, fastening with ropes not properly done; limit switches for rope travel control not working /not set properly;
- Steel slings failure on load, damaged due to aging/repeated overloading; wrong size usage may lead to accidents;
- Check electrical earthing connections, must be dependable;
- Over loading of main hook causing tardy movement of hoist;

(2) Gantry Cranes (moving on ground based rails):

***Operating problems are by and large similar to those listed above for EOT Cranes;**

***Additionally:**

- Ground based rails may develop misalignment due to foundation subsidence;
- Since in most cases these work outdoors, care has to be taken to protect the electrical installations and the motors from rain/ snow/dust; gantry structures also need anti-corrosive painting periodically;
- Trailing cables may develop snags due to exposure to atmosphere, to be checked periodically;

(3) Mobile Cranes (mostly Diesel Engine Operated, mounted on Auto-Vehicles):

- **Operating problems that can occur:**
  (a) Usual problems that are experience with Diesel Engines listed under paragraph **15.20.**
  (b) **Other problems** that can occur for the vehicle and the lifting beams and hooks:
    - Failure of the pneumatic tyres: punctures, erosions, air intake nipples developing snags/leaking;
    - Gear box problems: wear and tear on gear teeth, breakage on teeth, gear shifting lever eroded/broken, gear box bearings

worn out/jamming, low oil level causing accelerated erosions, loss of proper meshing and high level of noise;

- Hydraulic system developing snags: seal leakage, cylinders and rams corroded/bent, linkage mechanisms distorted/ displaced, pump malfunctioning(pump motor developing electrical faults, bearing failures, seals damaged due to aging,
- Lifting beam, hooks and steel ropes eroded due to aging and repeated over-loading;
- Operator's control levers and actuators developing snags due to erosions/aging;
- Torque/power transmission/clutch systems malfunctioning due to aging/material failures;
- Axle shaft and universal couplings failures due to wear and tear;
- Steering system developing snags due to wear and tear.

(4) Jib Cranes:

(5) Suspension Cranes:

Operating problems on these are generally as listed under:

- problems with hoists: electrical faults and bearings failures on motors, wear and tear on steel ropes and gears;
- electrical faults in pendant controller push buttons,
- electro-magnetic brake failures: coils burning out;
- **in case of Jib Cranes**, base gear and thrust bearing of vertical column developing snags due to wear and tear;
- unbalance on counter weights and failure/distortion of supports moving arm;
- **in case of suspension cranes,** roller wheels and their bearings eroded and causing erratic movements; flanges of tracks beams eroded due to long period operations;
- take care of greasing/lubrication points.

15.14 MATERIAL TRANSPORTATION EQUIPMENT

- **Referring to paragraphs-10.1.8 and 10.1.9 in Chapter-10, there are a variety of them being used in industrial units.**
- **Operating problems that are experienced with the following commonly deployed ones are listed out against each:**

(1) Motor Vehicles(Trucks, Trailers mainly):

Since these normally covered under Automobile Engineering Fields, these are NOT being dealt with here. It is expected that trained and experienced Engineers/Mechanics will be entrusted with their servicing , maintenance and repairs as and when required.

(2) Inter-Bay and Intra-Bay Electrically Driven Trolleys:

Operating problems that may require attention from time to time:

- Underground (placed in trenches) trolley lines developing electrical problems: faults on power supply cables, current collectors eroded loosing proper contact with bus bars/sparking and fire hazards, trailing cable faults,
- Over-ground rails getting misaligned/subside beyond permissible limits;
- Drive motor developing electrical and bearing faults;
- Gear box malfunctioning due to teeth erosions/breakage/lack of meshing, lack of lubrication/low oil level, coupling failures;
- Under-carriage structural failures: breakage/loosening of brackets and clamps/fixtures,
- Electrical faults in operator's console: contacts eroded, relays/contactors not working properly(coil burn outs);
- Regular over-loading causing faults on axles and gear box, occasional motor winding failures;
- Failures in braking system due to wear and tear: brake-fixture loosened, brake shoes eroded;
- Steel tyres getting misaligned due to failure of bearing and bearing housing fixtures;

(3) Wagons on Rails (for operation within Factory Premises only):

Many of the operating problems are similar to those listed above under paragraph-15.14 (2).

Since these are much bigger in size and capacity, these will have certain additional/unique operating problems as listed under, to be dealt with by trained experts:

***Suspension system related:**

- Laminated spring assembly loosened/a few leafs may get broken; support fixtures loosened/broken; bushes worn out/displaced;
- Lateral shift assembly (hydraulic/pneumatic) not functioning properly due to failures in the hydraulic/pneumatic sub-assemblies (defects in pumps, hoses, accumulator tanks, actuators);

***Other problems:**

- Buffers displaced/broken;
- Defects in coupling assembly;
- Rail tracks developing misalignments/subsidence at some places; rails eroded beyond permissible limits; track-switch-over fixtures/levers developing defects/jamming/lateral shift beyond permissible limits;

- Rail track holding down bolts loosened/broken; sleepers deteriorated due to aging; fish plate bolts loosened;
- Platform top surfaces damaged due to aging/regular loading-unloading;

(4) Rail Yard Shunters (Diesel Engine Driven):

- **Large factory Yards may require their services; these are not normally suitable for operating in Rail Station Yards(which require large capacity ones).**
- **Operating problems that are generally encountered can be presented in two parts: one related to the Diesel Engine; the other related to the shunter body/cabin and its under carriage parts/ fixtures;**

(4-1) Diesel Engine related:

(A) Preferred capacity is 75 HP Twin Engines: one for forward movement and the other for reverse movement; both start with dedicated battery-bank operated DC electrical cranking motors.

(B) **Operating problems may be as under**:

- Starting problems: battery-bank low on charge/low voltage; brush-gear and commutator developing electrical faults;
- Injectors getting choked, appropriate setting for fuel-air mixture not done or get disturbed in setting;
- Lubrication oil filters choked, oil has become old and dirty oil level depleted below permissible level; lub oil pipe leaking;
- Engine cylinder linings and piston rings eroded and not achieving proper compression, engine misfires;
- Fuel pump developing snags, fuel piping choked/leaking;
- Gear box developing problems of erosions/teeth broken/ jamming, Loss of meshing; spline shaft bearing failures, seals damaged and leaking; gear shifting levers eroded/bent/ displaced; oil level low and oil old and dirty;
- Clutch plates eroded and slipping , embedded springs weakened/broken;
- Operator's console in the cabin developing malfunctioning, linkage mechanisms distorted/displaced, electrical faults in the controller;

(4-2) Operating problems that can occur in the Under Carriage working systems:

- Universal coupling failing to transfer power to shaft and axles;
- Wheel bearing failure/jamming creating hot axle condition,
- Steel tyres wobbling due to loosened bearing housing;

- Lack of lubrication in the wheel bearings;
- Brake system failures due to aging/brake shoes worn out. Hydraulic-pneumatic pipes/hoses leaking/ruptured, vacuum pump not working properly, air tank leaking due to plumbing defects/ control valves not functioning properly,
- Wheel set fixtures loosened, not secured properly;
- Foundation fixtures of the engine can get loosened due to aging;
- There can be misalignments/distortions in the rail tracks due to aging (sleepers damaged, track holding down bolts and fishplate bolts loosened).
- Rail track subsidence at some places causing uneven movements;

15.15 WELDING MACHINES AND WELDING TECHNOLOGY

- **Readers are advised to refer to the brief on Welding Science and Technology presented at paragraph-4.4.1(3-1)(A-F) in Chapter 4.**
- **Welding machines of the following six types are commonly used in Industrial Units. Operating problems that are generally encountered are listed below against each type:**

15.15.1 In case of Gas Welding Machines

- Poor grade of gases being used may lead to un-even/unsteady flame causing lower temperature thus affecting quality of weld deposits;
- Inappropriate mixing of oxygen gas and the flaming gas(acetylene, propane, butane, hydrogen, natural gas, etc.) due to defects in the mixer-controller, clogging of nozzles of the flame torch, or wrong setting of gas flow rate: may lead to low flame/flame-out, distorted and sooty flame, also lower flame temperature;—all these may lead to poor quality of weld deposits;
- Excessive moisture in the gas cylinder /gas delivery pipes will affect weld quality;
- Hose pipes leaking/damaged due to aging;
- Thickness of work-piece is heavier than prescribed;

15.15.2 In case of Electrical Resistance Welding Machines

- Electrical faults in the transformer: primary input voltage low, in turn low voltage in the secondary causing low current flow into the weld point affecting weld quality;
- Terminals loose/eroded, cable connection not secure;
- Electrodes eroded, electrode pressure not enough,-pneumatic system malfunctioning (air pressure low);

- Control unit not working properly/electrical faults;
- Thickness of work-piece heavier than prescribed;
- Trailing cable damaged/insulation failure due to aging; undersized for the welding load current;
- Earth connection not proper causing erratic weld current flow.

15.15.3 In case of Electric Arc Welding Machines

- **There are Four types in use as presented below; operating problems that are faced are mentioned against each type:**

(1) AC Arc Welding Machines:

- Problems as listed above under 15.15.2 can occur in this case also ;

***Additionally:**

- Poor quality of electrodes/presence of excessive moisture in the electrode flux coating causing sputtering and un-even weld deposits;
- Wrong size of electrodes being used causing poor weld deposits;
- Earthing line/return line connections not secure causing disruption in the welding process;
- Current flow in to the arc point is low/transformer KVA rating is low; electrical faults in transformer winding causing power failures;
- Voltage and current setting in the transformer secondary not proper/ rotary switch contacts eroded due to aging/loose contacts;
- work-piece not prepared properly for the welding process;

(2) DC Arc Welding Machines:

- MG Set malfunctioning: electrical faults, commutator and brush gear developing short-circuiting/excessive surface erosions due friction and sparking (carbon brushes eroded/not set properly);
- **MG set** being a rotating machine with aging/prolonged running can develop bearing failures, insulation failures, damage/distortions on fans, excessive vibrations due to loosened foundation/ portable carriage structure not secure enough;
- Defective control unit/contacts eroded, defects on rotary switch;
- Trailing cables developing electrical faults/insulation failures;
- Overloading of generator due to heavier arc current flow due to use of heavier size electrode; long period continuous usage; under capacity MG Set being used;
- Work-piece not prepared properly for the welding process;

(3) **Submerged Arc Welding Sets:**

Beside the operating problems mentioned above at paragraph-15.15.3(1), the following problems may also occur:

- Malfunctioning of the feeding system of electrode wire from spools;
- Wrong grade/wrong size of electrode wire being used;
- Poor quality(excessive moisture/expired date)/wrong grade of welding flux powder being used;
- Flux powder feeding system developing malfunctioning/excessive moisture contents;
- Under capacity(low kva rating) machine not suitable for the selected work-piece;
- Work-piece not prepared properly for the welding process;

(4) **CO_2 Shielded Arc Welding Sets:**

Besides problems mentioned above at paragraph-15.15.3(1), the following problems may also occur:

- Supply of/rate of feed of CO_2 gas not appropriate/deficient in volume and pressure; supply pipe /tubes leaking; feeding system not working properly;
- Moisture content in gas beyond permissible limit, affecting quality of weld deposits(creating voids);

15.16 INDUSTRIAL X-RAY EQUIPMENT

Referring to the brief presented on the topic in paragraph-10.1.24(A) in Chapter-10, operating problems that may occur can be listed as under:

***Electrical System:**

- Supply voltage erratic,-voltage stabilizer not working properly;
- Transformer developing faults/insulation failures, faults in HT cables;
- Burn-out of X-Ray **tube,** positioning of X-Ray tube and target object not proper causing bad quality exposure/blurred images;
- Control console developing snags: contacts in rotary switches worn out, contactor coils burn-out,
- Exposure timer malfunctioning (over-exposure/under exposure);

***Others:**

- Over-sized target work-piece,
- Safety-shielding not conforming to prescribed system/configuration;
- Too frequent operation beyond recommended limits may reduce tube life drastically;
- Exposure-meter on operator need checking as per prescribed schedules;
- Poor quality of X-Ray plates/expired date ones being used;
- Processing of exposed plates not being done as per prescribed methods;
- Proper storage and retrieval system for exposed plates should be ensured;

15.17 INDUSTRIAL GAMMA-RAY EQUIPMENT

- **Referring to the brief on the topic in paragraph 10.1.24(B) in Chapter-10, operating problems be as under:**
  - Negligence in handling radio-active isotopes may create hazards for the operator and the immediate surrounding; all prescribed procedures must be adhered to;
  - Only trained persons must be deployed for operation and handling;
  - Radio-active isotope selection not proper.

15.18 MATERIAL TESTING EQUIPMENT

Referring to the brief on the topic presented a paragraph 10.1.25, in Chapter 10, operating problems may relate to:

- Malfunctioning of Electro-Hydraulic system: defects developing in hydraulic pump(electrical faults, bearing failures, seals damaged), leakages in pipes/hoses due to aging;
- Control apparatus and instrumentation developing defects mainly due to aging; actuators not working properly;
- Usual problems with electrical and electronic systems(cables faults, low voltage, defects in relays/contactors, failure of other circuit elements);
- Malfunctioning of associated computer(PC): network/software problems;

15.18.1 Gauges, Calipers, Micro Meters, Dial Gauges, Torque wrenches, RPM Counters;

These are sensitive precision devices and are not supposed to be serviced/ repaired by the user industry; with repeated use/aging, these are prone to loose accuracy; they are supposed to be got recalibrated from accredited Agencies/ Laboratories. It may be useful to enter into AMCs with the supplier/OEMs.

15.19 EARTH MOVING EQUIPMENT

***(Remark: with respect to those used in Factory Premises);* Referring to the brief on the topic presented in paragraph-10.1.9, in Chapter-10, operating problems that may occur:**

- Malfunctioning of Diesel Engines (drive): usual problems that can occur are similar to those listed in paragraph-15.20 below;
- Operating problems on Hydraulic/Electro-hydraulic systems ,by and large, are similar to those mentioned above under paragraphs-15.2.1(4-v) and 15.12.
- Power transmission system (clutch, gear box, gear shifting levers, gear box bearings) developing malfunctioning mainly due to wear and tear;

- Operator's console developing defects : failures of linkage mechanisms, electro-hydraulic actuators not working properly due to aging.

15.20 DIESEL ENGINES

REMARK: Diesel Engines that are used as Drives for Industrial (Machines; EXCLUDE Automobile Diesel Engines from this discussion.)

Referring to the brief on the topic presented at paragraph 10.2(8) (B) in Chapter-10, operating problems that may be encountered are briefly listed below:

(1) Engine Starting Problems:

(a) **Smaller Engines(hand cranking) up to 5 bhp:**

- Injector clogged, injector assembly leaking,
- Fuel pump not working properly, fuel filter clogged;
- Fuel pipe leaking,
- Cylinder/cylinder liner eroded due to aging, piston rings eroded/broken causing reduced compression pressure and misfire/knocking, sooty exhaust;

(b) **Multi-cylinder Engines upto 2000 bhp starting by Battery Bank powered Electrical (DC) Cranking Motor:**

> Beside the problems mentioned above:

- Battery bank charge low failing to provide adequate cranking power to the starting motor; electrical faults in charger unit, battery cells eroded;
- Cranking motor developing electrical faults, bearing/bush failure due to aging, armature burn out/short circuiting.

(c) **Multi-cylinder Engines above 2000 bhp with Compressed Air Starting**:

>Beside the problems mentioned above at (a):

- Compressed air pressure not enough to give required level of stating torque; air compressor not working properly, Air bottle valves leaking; air pipes leaking;
- Firing order not adjusted/primed according to prescribed setting;

(2) Problems of Running on Load (Multi-Cylinder Engines):

- Cooling water pump developing malfunctioning: electrical faults, bearing failure in pump motor, pump propeller eroded due to aging;
- Cooling water intake system failure/clogged; radiator clogged/ leaking, radiator fan not working properly: v-belt slipping/snapped, bearing failure;
- Malfunctioning of lubrication system: pump failure, electrical faults in pump motor, filter choked,

- Crank shaft bearings eroded beyond permissible limits due to wear and tear;
- Cylinder lining eroded, piston rings eroded/broken;
- Camshaft eroded, valve seats eroded causing uneven combustion timing/loss of combustion pressure; timing chain loosened due to wear and tear affecting firing order and engine knocking ;
- Speed governor not working properly, setting disturbed;
- Excessive vibrations due to uneven firing in cylinders/fly-wheel loosened, crank-shaft bearings (big-end bearings) loosened due to wear and tear;
- Low oil level in the sump and gear box;
- Load coupling bolts failure/loosened affecting torque transfer;
- Exhaust/silencer box choked/body leaking/muffler elements corroded/broken;

15.21 OXYGEN GAS PLANT

- **Referring to the brief on the topic presented in paragraph 10.1.19 in Chapter-10, operating problems that are generally encountered can be as under:**

(1) Air Compressor:

- Similar to those listed under paragraph-15.2.3 above;

(2) Air-Separation Unit:

- Pump dive motor developing electrical faults/ supply voltage low casing motor rpm slow down/bearing failures;
- Plunger pump developing snag: plunger rods eroded/seal leakage,
- Leakages in pipe lines and flanges/gasket damaged,
- Instruments not working properly due to aging ;
- Back pressure from O_2 and N_2 receiver/storage tanks: non-return valves not working properly,

(3) Storage and Distribution of Oxygen Gas:

- Oxygen gas being explosive in nature, good care is to be taken to avoid leakages from receiver tanks and distribution pipe lines/ valves;
- Pipe lines of O_2 and N_2 have to be properly colour-coded to avoid mix-up;
- Keep fire-protection system in active mode with periodical checks;

(4) Storage and Disposal of N_2 Gas:

N_2 being a by-product of the plant, proper arrangement has to be in place for safe and cost-effective storage and disposal;

15.22 ACETYLENE GAS PLANT

- **Referring to the brief on the topic presented at paragraph-10.1.20 in Chapter-10, the operating problems that are generally experienced can be listed as under:**
  - Acetylene gas being highly combustible, all prescribed safety measures must be taken, negligence can lead to fire hazards/explosion;
  - Regular checks on leakages from the reactor chamber, pipe lines and cylinders must be done;
  - Choking of gas generator chamber due to excessive accumulation of slag;
  - Poor quality and wrong size of calcium carbide lumps can reduce gas yield;
  - Control system of feeding calcium carbide lumps and water dosing require regular checks for ensuring safe operation;
  - Disposal of slurries being corrosive, have to be done with due care;

15.23 PRODUCER GAS (COAL GAS) PLANT

- **Referring to the brief on the topic presented in paragraph 10.1.21 in Chapter-10, operating problems generally encountered can be listed as under:**
  - Low gas yield due to poor quality/wrong grade/lump size of coal being used;
  - Drive system of the rotary hearth(motor, gears, linkage mechanisms) developing malfunctioning (electrical faults, bearing failures, gear teeth slipping/poor meshing, low supply voltage);
  - Defects in control panel, electrical/mechanical faults in switch gears operating mechanisms;
  - Malfunctioning developing in the Electrostatic Precipitators (electrical faults/insulation failure in the HV-DC system, structural failure in the bus bars/ failure of ceramic insulators; precipitator chamber clogged with coal tar/bituminous deposits;
  - Excessive clinker/slag generation due to malfunctioning/over loading of rotary hearth;
  - Defects developing in oil/gas burners(clogging/disturbed setting of nozzles;
  - **Water spraying system** malfunctioning/nozzles clogged, water pressure low due to defects in pump; water treatment plant not working properly;
  - Regular attention has to be paid to the safe storage and disposal of generated coal tar/bitumen and phenolic liquids(hazardous);

- Regular attention has to be paid to the safety of the burn-out system for the surplus gas(which cannot be discharged to the atmosphere).
- Regular attention is required to avoid/effectively control within safe limits, gas leakages from the distribution pipe lines and the consuming points; colour-coding done correctly;

15.24 TROUBLE SHOOTING ON PORTABLE HAND TOOLS AND WORK-SHOP DEVICES

(1) In case of Electrically Operated Hand Tools:

- **These hand-held tools are normally operated by fraction horse power single phase Universal (AC/DC) series commutator motors.**

  Operating problems generally experienced are as under:
  - Electrical faults in cables/ motor/control switches; supply voltage low;
  - Bearing/bush worn out causing excessive vibration and lower speeds;
  - Abuse by operators leading to armature burn outs;
  - Gears (mainly worm gears) worn out/teeth broken;
  - Sparking at brush tips on the commutator surface;
  - Washers/separator plates/baffle plates worn out/damaged/bent;
  - Field coils damaged/insulation failure;
  - Tool grip mechanism defective/worn out; tool piece worn out;

(2) In case of Pneumatic/Compressed Air operated Hand Tools:

- Low air pressure in the line;
- Hoses leaking/damaged ;
- Wear and tear on bearings/bushes; washers/seals/lock nut;
- Rotor vanes damaged/bent; spindle coupling damaged,
- Tool grip mechanism defective; tool piece defective;

(3) In case Hydraulically Operated Portable Tools:

- Failure of hydraulic pressure system: electrical faults, bearing failures in pump motor; trailing cable damaged/shorted;
- Cylinder and piston eroded/seals leakage/hoses damaged and hose coupling leaking; poor quality of hoses used;
- Mechanical linkage/lever system malfunctioning due to aging;
- Quality of hydraulic fluid used not appropriate for the device;
- Electro-hydraulic actuators defective, not working properly; plungers jamming;

(4) In case of Petrol Engine Driven Portable Tools:
(Such as Chainsaws, Timber saws, Rock-cutters/Drillers);

- Air-cooled small petrol engines : ignition failures/spark-plug not working properly/ clogged, body earth leakage,
- Fuel injection system clogged causing engine misfire;
- Engine stop running due to over-heating being run for longer duration than recommended;
- Cutter chains/drills/ blunted , getting stuck-up on work-piece; broken due to over-loading;
- Chips/dust from work-piece clogging the engine exhaust line causing stoppage
- Usual wear and tear on moving parts which may lead to engine failure.

■ ■ ■

Annexure 1: Quantities and Units

The term quantities designates physical and technical materials characteristics of objects and processes. Base quantities physically independent of one another.

Base Units of the international System of Units (SI)
(As defined in "Law of Units in Measurement")

Quantity	Unit	Symbol
Length	Meter	M
Mass	Kilogram	Kg
Time	Second	S
Electric current	Ampere	A
Thermodynamic Temperature	Kelvin	K
Amount of Substance	Mole	Mol
Luminous Intensity	Candela	Cd

Definition of Base Units (Considerably Simplified in Part)

1 Meter is the distance covered by light in a vacuum in 1/299792458 second.

1 Kilogram equals the mass of the international prototype Kilogram.

1 Second equals 9,192,631770 periods of the radiation of an atom of **Cesium-133**.

1 Ampere is the intensity of a current which, by flowing through two indefinitely long, parallel current-carrying conductors, exerts a force of 2/10 Newton/m upon this pair of conductors.

1 Kelvin is equal to 1/273. 16 of the absolute temperature of the triple point of water

Contd...

1 Mole is the amount of substance of a system which contains as many elementary entities as there are atoms in 0.0112 kg of **Carbon-12.**

1 Candela is the luminous intensity in a given direction of a radiant source which emits radiation of a frequency of $540x10^{12}$ hertz and has a radiant intensity in that direction of 1/683 watt per unit solid angle.

Selected Derived Units and Terms

Differences in temperature are expressed in Kelvin Temperature in degrees Celsius/°C equals thermodynamic temperature/K-273.15

Temperature range 1°C equals 1 temperature range 1 K.

Temperature in degrees Fahrenheit/° F equals 9/5 (temperature in degree Celsius/° C)+32; 1.8° F equals 1° C;

Name of Unit	Unit symbol	Definition	Previous unit Symbols
Becquerel	Bq	1 Bq = 1/s	1 Ci = 37 G bq
I Hertz	Hz	1 Hz = 1/s	
Joule	J	1 J = 1 N x m	
Newton	N	1 N = 1 kg x m/s^2	
Pascal	Pa	1 Pa = 1 N /m^2	
Siever	Sv	1 Sv = 1 J /kg	1 rem = 0.01 Sv
Steradiant	Sr	1 Sr = 1 m^2 m^2	
Watt	W	1 W = 1 J/s	

Conversion of British and American units to Metric Units

Units of length	m	in	ft	yd	Stat. mile	n. mile
1 m	1	39.3701	3.28084	1.09361	0.00062	0.00054
1 inch	0.0254	1	0.08333	0.02778	–	–
1 foot	0.3048	12	1	0.3333	0.000189	–
1 yard	0.9144	36	3	1	0.0005688	–
1 statute mile	1609.3	63360	5280	1760	1	0.868976
1 n mile	1852	72960	6076.12	2025.12	1.15078	1

Yard, British unit 1 a = 1 are = 100m^2 1 US gallon = 0.0037854m^2
1 statue mile = 1 Land mile 1 ha = 1 hectare = 1000m^2
1 UK gallon = 0.0045461 m^3
1 n mile = 1 nautical mile; 1 acre = 4,046.86 m^2 1 Barrel petroleum = 0.158971 m^3
1 fathom = 6 ft = 1.8288m 1 US fluid ounce = 1.85m^3
1 UK fluid ounce= 1.174 m^3

Contd...

Units of Mass

Units of Mass Weight	Kg	T	Oz	Lb	Sh cwt	Cwt	Sh tn	Ton
1 kg	1	0.001	35.274	2.20462	–	–	–	–
1 t	1000	1	35274	2204.62	22.0462	19.685	1.10231	0.98421
1 oz	0.028535	–	1	0.0625	–	–	–	–
1 lb	0.45359	–	16	1	0.01	0.0089	0.0005	–
1 sh cwt	45.3592	–	–	100	1	0.8929	0.05	0.04464
1 cwt	50.8023	–	–	112	1.12	1	0.056	0.05
1 sh tn	907.185	–	–	2000	20	17.857	1	0.9929
1 ton	1016.05	1.01605	–	2240	22.4	20	1.12	1

T = metric ton = tonne 1 sh cwt = 1 short hundred weight,(US unit)= ton (British Unit)

1 oz = 1 ounce avoirdupois 1 cwt = 1 hundred weight, (British unit)

1 lb = 1 pound avoirdupois 1 short ton (US unit)

Units of Force	N	Dyne	P	Kp	Lbf
1 N	1	105	101.9716	0.10119716	0.224809
1 dyne	10^{-5}	1	1.01976×10^{-3}	1.019710×10^{-6}	$2.248\text{-}9\times 10^{-5}$
1 p	9.80665×10^{-3}	980.665	1	0.001	2.20462×10^{-3}
1 kp	9.8o665	$9.8\text{-}65 \times 10^{5}$	1000	1	2.20462
1 lbf	4.44822	4.44822×10^{5}	453.592	0.453592	1

Dyne (unit of force in the centimeter-gram-second system) 1 kp = 1 kilopound

1 p = 1 pound 1 lbf = 1 pound –force = 1 lb × 9.81 m/s^2

Units of Pressure	Pisa	Bar	At	arm	Torr	Psig
1 P1	1	10^5	$1{,}01976 \times 10^{-5}$	$0{,}98692 \times 10^{-2}$	$0{,}750062 \times 10^{-3}$	$145{,}038 \times 10^{-8}$
A bar	10^5	1	1,019716	0,986923	750,062	14,5038
1 at	$0{,}980665\times10^5$	0,980665	1	0,967841	735,559	14,2233
1atm	101,325	1,01325	1,033327	14	760	14,69596
1 Torr	133,3224	$1{,}333224 \times 10^3$	$1{,}315789 \times 10^{-3}$	$1{,}315789 \times 10^{-3}$	1	$19{,}3368 \times 10^{-3}$
1 psi	$6{,}89476 \times 10^{-5}$	$68{,}947 \times 10^{-3}$	703070×10^{-3}	$68{,}0460 \times 10^{-3}$	51,7128	1

1 bar = 108 dyne/cm2 Psia = Pounds per square inch absolute

1 at = 1 kp/cm2 = 10mWC(technical atmosphere)

psig = pounds per square inch ,gauge

1 atm = 760 torr(standard atmosphere)

psid = pounds per square inch differential

1 psi = 1 lbf/in2 (pound-force per sq. inch)

Contd...

Units of Energy	J	kWh	PSh	kpm	Kcal	Btu	SKE
1J=1Ws	1	$2{,}778\text{x}10^{-7}$	$3{,}77\text{x}10^{-7}$	1.1019716	$2.388\text{x}10^{-4}$	$9.478\text{x}10^{-4}$	34.12×10^{-4}
1kWh	3.6×10^{6}	1	1,35962	$3.671\text{x}10^{5}$	859.845	3,412.14	12.28×10^{-2}
1 PSh	2.64×106	0,735499	1	$2.7\text{x}10^{5}$	632.41	2,509.62	90.36×10^{-3}
1kpm	9,800665	$2{,}724\times10^{-6}$	$3{,}70.10^{-6}$	1	2.342×10^{-3}	9.295×10^{-3}	33.47×10^{-8}
1kcal	4,186.8	$1{,}163\times10^{-3}$	1.581×10^{-6}	426.935	1	3.96832	14.29×10^{-5}
1Btu	1,055.06	$2{,}931\times10^{-4}$	3.985×10^{6}	107.586	0.251996	1	35.99×10^{-6}
1 SkE	$29{,}307\times10^{-6}$	8.141	11,067	2.988×10^{6}	7000	27.78×10^{6}	1

1 kWe = 1 kilowatt-hour

2 BTU = 1 British Thermal Unit

1 kcal = 1 kilocalorie

1 kpm = 1 kilopound-meter

1 PSh = PS-hour

1 SKE = 1 CE (coal equivalent)

Units of Power	KW	PS	hp	Kpm/s	Kcal/s	Btu/s	Ft-lbf/s
1 KW	1	1.35962	1.34102	101.9716	0.238846	0.94781	737.562
1 PS	0.735499	1	0.986320	75	0.1757	0.69712	542.476
1 hp	0.745700	1.01387	1	76.042	0.17811	0.70679	550
1kpm/s	9.80×10^{-3}	0.013333	0.0131509	1	2.342×10^{-3}	9.295×10^{-3}	7.23301
1 kcal/s	4.1868	5.692	5.614	426.939	1	3.96832	3.088.05
1 Btu/s	1.05505	1.4345	1.4149	107.586	0.251993	1	778.17
1 ft-lbf/s	1.356×10^{-3}	1.843×10^{-3}	1.818×10^{-3}	0.138255	3.238×10^{-4}	3.285×10^{-3}	1

1 KW = 1 kilowatt = 10 10 erg/s= 1 kj/s

1 kpm/s = 1 kilopoundmeter per sec.

1 Btu/s = 1 British thermal Unit/sec

1 kcal/s = 1 kilokalorie per sec.

1 PS = 1 metric horsepower

1 ft-lbf/s = 1 foot-pound / force/sec

1 hp = 1 horsepower

Source: Siemens AG Pocket Diary, 1998.

■ ■ ■

Annexure 2: Energy Terms and Definitions

Coal: Is the most abundant fossil primary energy source and the classic fuel for power and heat generation. The importance of coal will be further increased in the future by gasification and liquefaction processes.

Energy: is defined as the capacity of a system to produce an external effect (work, heat, light). As a matter of principle of physics, energy can only be converted from one form of energy into another. For electrical energy, the unit of measure is the kilowatt-hour (symbol: kWh).

Coal Equivalent (CE): Is the reference standard for the energy content of various fuels. 1 CE or 1 kg CE represents the mean energy content of 1 kg of hard coal having a calorific value of 7,000 kcal. 1 TCE=29.3 GJ.

Combined-Cycle Power Plants: The combination of gas turbines and steam turbines in fossil-fired power plants provides advantages in respect of short construction periods, financing, flexibility, economic efficiency and environmental compatibility both where new projects or re-trofitting is concerned.

Combined Heat and Power (Cogeneration): The simultaneous supply of electricity and process heat or district heat by a power plant. Utilization factors of more than 80% can be achieved.

Efficiency: Efficiency is a measure of effectiveness of an energy conversion process, i.e. the ratio of the useful energy output to the total energy input.

Power: Inherent or induced capacity of a system (drives, gadgets, plants etc.) to do measurable work (or deliver force to do work over a period of time to unleash energy).

Electric Energy: Is defined as the electric power used over a specific period of time. If a 100-watt electric bulb burns for ten hours, the electric energy consumed is 1 kWh. (power x time=energy consumed).

Emissions: Are defined as releases of a pollutant into the environment, e.g. gases, dust , waste heat , noise, vibration.

Energy: Is defined as the capacity of a system to produce an external effect (work, heat, light). As a matter of principle of physics, energy can only be converted from one form of energy into another. For electrical energy, the unit of measure is the kilowatt-hour (symbol: kWh).

Energy Programme: Goals set by local authorities, an industrial enterprise or a government to assure the supply of energy as well as measures to be taken to achieve these goals.

Energy Recovery: Utilization of energy remaining (waste or residual) as a by-product, e.g. of production processes, ventilation and air conditioning or in households. Frequently , it is technically feasible and attractive from the economic point of view.

Forms of Energy: In the scientific sense, a distinction is made, for example, between mechanical, electrical, chemical and magnetic energy, heat and nuclear energy and in the practical sense between primary energy, derived energy, final energy and useful energy.

Gas Turbine: A gas turbine is a device which involves fresh-air compression, hot-gas production and the generation of rotational movement in one casing. (basically a IC-Engine)

Generator: The mechanical energy of a drive shaft is converted to electrical energy in a generator using the dynamo-electric principle.

Gross Calorific Value: The gross calorific value of fuels indicates the energy liberated during complete combustion of a unit of mass of a fuel and the recovery of water vapour condensation heat.

Petroleum: Is a naturally occurring mixture of low or high viscosity liquid consisting mainly of hydrocarbon derivates. Through selected distillation and cracking processes in the petroleum refinery, it is separated into fractions by boiling range, density and viscosity; it is a fossil fuel extracted from liquid gas reservoirs.

Quality of Power and Energy: In the energy and power plant sector is expressed in terms of time, function and cost (period of availability of stable supply, reliability of supply at prescribed parameters, cost of supply etc.).

Rated output: The rated output of a power plant is the maximum continuous rating for which the generators have been designed.

Wind Energy: The energy derived from wind flow turning a propeller driver wind turbines which rotates a coupled generator for generation of electric power in a localized load center; large scale generation not feasible at one location.

Solar Energy: Predominantly originates from thermo-nuclear fusion reaction going on within the sun. Commercial solar energy utilization is made difficult by a low energy density and a strongly fluctuating insolation time. Presently, solar energy is converted directly to electrical energy by using photo-voltaic plates exposed to sun rays.

Sources of Energy: Are defined as materials or systems that contain useful energy, e.g. hard coal, uranium, water, fossil, gas/ fuels, solar rays, wind flow, falling water, burning wood.

Instrumentation and Control: In power plants measurements, control and monitoring are known collectively as instrumentation and control.

Load Ranges: The time dependent power demands of the ultimate loads on the system require a power plant operating mode adapting in terms of time. Depending on their operational and economic characteristics, power plants are operated for 24 hours, for a number of hours at a time, or merely for a short time during the day, i.e. base-load, intermittent load and peak-load operations.

Maximum Capacity: The maximum capacity of a power plant is defined as the maximum continuous capacity, limited by the lowest capacity part or system, and is determined by measurements during operation.

Natural Gas: Is the fossil primary energy source extracted from the underground hydrocarbon fields with the highest increases in demand. Besides the economic aspects, environmental protection favours the utilization of natural gas in power generation.

Nuclear Power Plants: Are thermal power stations utilizing the nuclear fuels (uranium, plutonium and thorium) with releasing combustion gases. Heat generated in the reactor due to atomic fission process is utilized in a HRSG to generate steam which drives steam turbines which in turn drives a electric power generator.

Oil Equivalent (OE): Is the referenced standard for the energy content of various fuels. 1 kg OE represent the mean energy content of 1 kg of oil having a calorific value of 10,000 kcal.

$$1 \text{ TOE} = 41.9 \text{ GJ}.$$

Steam Turbine: This is a thermal machine converts the thermo-dynamic energy contained in steam to mechanical energy in the form of rotor revolution.

Usage Factor: The usage factor of power plants is defined as the average utilization of the maximum capacity within a specific time, measured in hours per annum (h/a) .

Plant Load Factor (PLF): This is defined as the level of capacity utilization of a power plant with respect to its designed/installed capacity.

Hydro-Turbine: This is a machine which converts the kinetic energy of falling water into mechanical rotating energy which in turn drives a generator for generating electric power.

Utilization Factor: The utilization factor of a power plant is defined as the average operating efficiency, taking into consideration the part load behavior.

■ ■ ■

Annexure 3: National Laboratories and Institutes (A selected listing)

The following Laboratories and Institutes in India are doing fundamental and applied research on the subject mentioned against each. Industry, trade, other research organizations, and even educational institutions can benefit from their research and application of scientific knowledge.

Name	Location	Functions performed
1. National Physical Laboratory	New Delhi	Research in problems relating to physics, both fundamental and applied. Maintenance of Standard Testing and Measuring facilities are also available.
2. National Chemical laboratory	Pune	Fundamental and applied research covering the whole field of Chemistry for which other specialized institutes have not been set up. The National Collection of Type Cultures is housed in the Laboratory.
3. National Metallurgical Institute	Jamshedpur	Fundamentals and applied metallurgical research. It also maintains Regional Laboratory for Foundry Research Stations.
4. Central Drug Research Institute	Lucknow	All aspects of drug research including evaluation and standardization of crude drugs, Discovery of substitutes for Pharmacological and synthetic chemical biochemistry and bio-physics, infection immunization Pharmacology, Chemotherapy and experimental medicine.
5. Central Food Technological Research institute	Mysore	Food processing and conservation of food, food engineering and all aspects of fruit technology. Regional fruit and vegetable preservation stations at Trichur, Nagar, Mumbai, Shimla and Lucknow have been established.
6. Central Fuel Research Institute	Dhanbad	Fundamental and applied research on fuels-solid, liquid and gaseous. Physical and chemical surveys of Indian coals are conducted through seven coal survey stations under the institute.

Name	Location	Functions performed
7. Central Glass and Ceramic Research Institute	Jadavpur (Kolkata)	Research on different aspects of glass and ceramics, pottery, porcelain, refractory and enamels; development of processes for manufacture of glass and ceramic articles; standardization of raw materials used in the ceramic industry.
8. Central Building Research Institute.	Roorkee	Engineering and structural aspects of buildings and human comforts in relation to buildings.
9. Central Electronics Engineering Research institute.	Pilani (Rajasthan)	Design and construction of electronic equipment and components and testing equipment.
10. Central Electro-Chemical Research institute	Karaikudi (Tamilnadu)	Research on different aspects of electro-chemistry, including electro-metallurgy, electro-deposition and allied problems.
11. Central Leather Research Institute	Chennai	Fundamental and applied aspects of leather technology. It has regional extension centres at Mumbai, Kolkata, Kanpur, Rajkot and Jalandhar.
12. Central Mining Research Station	Dhanbad	Research in methods if mining, Research on safety in mining and mine machinery
13. Regional Research Laboratory	Hyderabad	Research in problems relating to the industries and raw materials of the region.
14 Indian institute of Experimental Medicine	Kolkata	Research in various aspects of biochemistry, as applied to medicine bacteriology etc.
15. Central Salt and marine Chemical Research Institute	Bhavnagar (Gujarat)	Investigation on production of pure salts; reduction in cost of production, economic by-products of salt manufacture; of new methods and techniques for the recovery, production and utilization of marine and allied chemicals.
16. Earthquake Research Institute	Roorkee	Research on seismic zones and development of technology for earthquake damage control systems etc.
17. Hydrology Research Institute	Roorkee	Hydrological research, river research, water shed research etc.
18. Regional Research Laboratoy	Jammu Tawi (J&K)	Research in problems relating to the industries and raw materials of the region and research specially directed to medicinal plants of the Himalayas (Kashmir Region).
19. Central Mechanical Engineering Institute	Durgapur (West Bengal)	Research in Mechanical Engineering in all aspects.
20. National Environment Engineering Research institute	Nagpur (Maharashtra)	Research in all aspects of public health and engineering, coordination of work of all interested agencies in this field in the country.
21 National Aeronautical Laboratory	Bangalore	Scientific investigation of the flight problems with a view to their practical application to the design, construction and operation of aircraft in India.
22. Indian Institute of Petroleum	Dehradun (Uttarakhand)	Research in petroleum refining and processing of natural gas petro-chemicals providing facilities for the training of personnel for petroleum industry.
23. National Geophysical Research Institute	Hyderabad	Correlation of the field data in all fundamental aspects of geology and geophysics with laboratory investigations and theoretical studies.

Name	Location	Functions performed
24. National Institute of Oceanography	Panaji	Research on various aspects of physical, biological, geological and chemical oceanography including prospecting for petroleum and minerals in sea bed.
25. Regional Research Laboratory	Jorhat (Assam)	Research in problems relating to more efficient utilization and better conservation of important national resources of Assam and other regional needs posing special problems.
26. Central Indian Medicinal Plants Organization	Lucknow	Co-ordination of activities in the development of cultivation and utilization of medicinal plants on organized basis.
27. Central Scientific Instruments Organization	Chandigarh	Promotion and development of indigenous manufacture of scientific instruments of teaching ,research and industry
28. Indian National Scientific Documentation Centre.	New Delhi	Provides full range of documentation services.
29. National Register unit	New Delhi	Maintenance of information pertaining to Indian scientific and technical personnel in the country and abroad.
30. Central Design and Engineering Research institute	New Delhi	Provides assistance in the translation of laboratory results into industrial practice and designing of plants based on process developed.
31. Regional Research Laboratory	Bhubaneshwar	Research in problems relating to the industries and raw materials of the region.
32. National Biological Laboratory	New Delhi	Fundamental and applied research in modern of biological science.
33. Structural Engineering Research Centre.	Roorkee	Research in specialized design and development work in structural problems connected with buildings, bridges and other structures.
34. Industrial Toxicological Research Centre	Lucknow	Studies in harmful effects of industrial toxins on skin, blood, Gastro-intestinal tract, central nervous system, bones etc.
35. Bhabha Atomic Research Centre (BARC)	Trombay (Mumbai)	Atomic and Nuclear energy Research centre.
36. Tata institute of Fundamental Research.	Mumbai	Scientific research institute (fundamental research in all branches of sciences).
37. Institute of Plasma Research	Gandhinagar (Gujarat)	Scientific research on plasma physics and controlled nuclear fusion technology.

■ ■ ■

Annexure 4: Some Important Industrial/ Business Directories

1. Indian Industry Directory;
2. The Indian Business Directory;
3. Directory of 15000 Indian Manufacturers, Exporters and Suppliers;
4. CII (Confederation of Indian Industry) Directory;
5. World Industry and Trade Organizations in 200 Countries;
6. Handbook of Research Institutions in India;
7. The Gujarat Directory of Manufacturers;
8. Kothari's Industrial Directory of India;
9. World Importers' Directory;
10. Kompas India;
11. Directory of useful Addresses;
12. Indian Pharmaceutical Guide;
13. Hotels and Restaurants Guide of India;
14. Diary India;
15. Indian Poultry Industry Yearbook;
16. Indian Electronics Directory;
17. Drugs Directory;
18. All India Textiles Directory;
19. Gujarat Plastic Manufacturers Association Directory;
20. TATA Yellow Pages;
21. Directory of Indian Processed Food and Allied Industries;
22. Chemical Weekly Buyers Guide;
23. All India Ready-made Garments Exporters Guide;

24. Rubber and Rubber Goods Industry in India;
25. All India Directory of Woolens, Silk, and Textile Industries;
26. Textile Directory of Ahmedabad;
27. Indian Businessman's Guides to Americas, Africa, Japan, South East Asia; Gulf Countries, Commonwealth Countries;
28. Apeda Export Statistics For Agro and Food Products of India.

NOTE: (i) In addition to the above listing, one may refer to telephone Directories, the yellow pages, and other such ready reference sources.

(ii) The above directories and hand books are available in large book stalls.

■ ■ ■

Annexure RB: List of Books for Supplementary Reading

Sl. No.	Title of the Book	Author/Publisher
1.	Mechanical Engineering	R.S. Khurmi, J.K. Gupta, S. Chand Publishing, New Delhi.
2.	Objective Mechanical Engineering	P.K. Mishra, Upkar Publishing.
3.	Mechanical Engineering Systems	Richard Gentle,
4.	Theory Of Machines	J.K.Gupta, R,S Khurmi,
5.	Design of Machine Elements	V.B.Bhandary.
6.	Machineries Handbook	Erik Oberg, Amazon.
7.	Modern Machine Shop Practices (Vol-I & II).	Joshua Rose, Amazon.
8.	Work Shop Practices	G.S. Bawa, K.S. Yadav, Swaran Singh.
9.	Work Shop Technology (Part-I, II, III.)	Dr. WAJ Chapman, Amazon;
10.	Work Shop Technology	G.S. Bawa, Pusa Polytechnic.
11.	Text Book of Refrigeration & Air Conditioning	R.S.Khurmi.
12.	Mechanical Technology,	R.S.Khurmi.
13.	Engineering Thermodynamics	P.K. Nag.
14.	Kinematics of Machines	Sadhu Singh.
15.	Clean Machining	Djebara Abdelhakim.
16.	Mechanisms and Machine Theories	J.S.Rao.
17.	Fundamentals of Machining and Machine tools	R.K.Singhal;
18.	Practical Boiler Operation Engineering and Power Plants	PHI- Learning/Publishing.
19.	Welding: Principles and Applications	Larry F Jeffus.

Contd...

Sl. No.	Title of the Book	Author/Publisher
20.	Pumps: Principles and Practices	Hubbert Edwin Collins.
21.	Handbook Of Mechanical Engineering	McGraw Hill ,USA.
22.	Handbook Of Industrial Engineering	McGraw Hill, USA.
23.	Tool Engineers Handbook	McGraw Hill, USA.
25.	Foundry Technology	O.P. Khanna
26.	Hydraulic and Pneumatics	E.A.Parr.
27.	Principle of Foundry Technology	P.L.Jain.
28.	Materials And Supply Chain Management	D.J. Syam.
29.	Handbook Of Business Administration Practices	D.J.Syam.
30.	Capital Investment Projects	D.J.Syam.
31.	Finance for Non-Finance Executives	N.J.Yasaw.
32.	Business Economics	Dr. Deepa Shree.
33.	Operations Management	William R Stevenson.
34.	Management Guide for Engineers and Technical Persons	McGraw Hill, USA.
35.	Material Handling: Traffic and Transportation	Elias S Tyler.
36.	Fluid Mechanics and Hydraulic Machines	R.Kumar.
37.	Production and Operation Management	Gaither Newman.
38.	Enterprise Resource Management	D.J.Syam.
39.	Organization and Management	B.S.Moshal.
40.	People Who Make Profit	S.K.Bhatia.

■ ■ ■

Printed in the United States
by Baker & Taylor Publisher Services